Vascular Organization of Angiosperms
A New Vision

Vascular Organization of Angiosperms

A New Vision

Jean-Pierre André
Institut National de la Recherche Agronomique
Antibes
France

Science Publishers, Inc.
Enfield (NH), USA   Plymouth, UK

SCIENCE PUBLISHERS, INC.
Post Office Box 699
Enfield, New Hampshire 03748
United States of America

Internet site: *http://www.scipub.net*

sales@scipub.net (marketing department)
editor@scipub.net (editorial department)
info@scipub.net (for all other enquiries)

Library of Congress Cataloging-in-Publication Data

André, Jean-Pierre.
 [Organisation vasculaire des angiosperms. English]
 Vascular organization of angiosperms: a new vision/Jean-Pierre André.
   p. cm.
 Includes bibliographical references.
 ISBN 1-57808-310-9 HC/1-57808-382-6 PB
  1. Vascular system of plants. I. Title.

QK725.A557 2005
575.7 — dc22

                                                    2005044055

ISBN 1-57808-310-9 HC
ISBN 1-57808-382-6 PB

Published by arrangement with INRA, Paris.

Translation of: ***Organisation Vasculaire des Angiospermes:***
                Une Vision Nouvelle
French edition: © INRA, Paris, 2002.
                ISBN 2-7380-0995-6

Published by Science Publishers, Inc. Enfield, NH, USA.
Printed in India.

To the memory of Louis Gachon, former head of the agronomic lab in Clermont-Ferrand, then of the Agronomic Department of INRA, who, until his death, encouraged and supported my research with his loyal and friendly kindness.

Foreword

This work by Jean-Pierre André will be appreciated at several levels. Some readers may simply leaf through it as if it were an art book. Biologists will discover in it an unexpected aspect of vascular structures of plants, while others might mistake some of the images for a mountain relief, the tentacle of an octopus, a crab's leg, or the columns of an Egyptian temple. But anyone who picks up this book will be seduced, as I myself was when I first discovered them under the microscope, by the diversity of forms and surfaces of the microcasts made by the author. More important, the book provides readers with clear explanations and elegant and precise illustrations, which allow them to explore in an innovative manner and better understand the spatial organization of the vascular system of plants. The pedagogical preoccupation that underlies the work will be appreciated by educators as well as students.

But this book goes much further. It reveals the existence of vascular structures that have till now remained unknown. It thus challenges some of our ideas about the circulation of raw sap and, therefore, the relationships between the organs of a single plant or between a parasite and its host. The microcasting technique, apart from its obvious interest for systematists and wood technicians, invites us to venture into new areas of research in ecology and physiology. The images obtained also raise the question of the control of differentiation of the vascular edifice. Wood is often considered a near perfect example, the paradigm, of plant differentiation since that differentiation leads to the irreversible death of conducting elements, while a large number of plant cells considered differentiated can in certain conditions return to a juvenile state. The complexity of the architecture revealed by microcasts is indeed quite far from simple systems such as those of Zinnia cells that presently serve as models for research on the differentiation of the xylem.

Finally, the applications of microcasts are not limited to the study of vascular organization. Part I of the work contains a section presenting some potential applications, such as the exploration of secretory structures or of intercellular spaces. Part II gives us the necessary technical information to undertake our own explorations.

Not only does this book allow us to appreciate the pioneer work accomplished by Jean-Pierre André over the last ten years, it also opens the door to new investigations in various directions. It is a work that will be useful to wood biologists as well as to educators and researchers. We wish the author all the success he deserves.

Anne-Marie Catesson

Preface and Acknowledgements

Ever since my colleague Albertine Champéroux, amid rows of rose bushes, initiated me into the fundamentals of histogenesis and plant organization, it has been for me a decade of passionate discovery of plant anatomy "through the half-open window of microcasting", after about thirty years working in other lines of research.

During this time, I was forced to perfect an original technique to study vessels, in the absence of sufficiently effective knowledge and mastery of the conventional histological techniques, and then to test it under highly varied conditions, in the hope of some original "lucky finds" in the field of vascular anatomy. For the freedom to pursue this research, I express my gratitude first of all to the Institut National de la Recherche Agronomique.

From the beginning, many people have given me the benefit of their assistance and their knowledge: I am pleased to acknowledge their help and testify to my active remembrance of them.

Since the first attempts at vessel casting in 1992, my colleague and friend Pierre Cruiziat expressed a sustained interest in the development and implementation of the technique, owing to his own researches in the circulation of water in the plant, and stimulated me by his curiosity and encouragement until the publication of this work. I owe a good part of the success of this development to the advice of René David, a specialist in archaeological casting, with whom I came into contact through a news article.

In Anne-Marie Catesson, I have had the good fortune to find the Master to whom I owe the essence of my deeper knowledge of plant anatomy and histology. Ever since my first visit to the Ecole Normale Supérieure, she has shown her interest in this unusual technique and given me her confidence and patient support. We collaborated in the study of the "heterogeneous vessels" revealed by casting in the abscission zones.

All the plants examined were provided voluntarily, the Solanaceae and bamboos by helpful regional producers, the cereals by the INRA laboratories, the ligneous species by the Jardin Thuret at Antibes, and the Guyanese

Pteridophytes by distant friends. The rest was obtained from the surrounding countryside.

I am grateful to Danielle Repoux, of the Ecole des Mines at Sophia Antipolis, and Lucien Roger, of INRA at Montpellier, who succeeded her, for having provided with competence and patience more than 150 hours of scanning electron microscopy facilities and produced more than 1500 images. All these images were then passed on to Christian Slagmulder, of INRA at Antibes. If the sumptuous, unanimously admired enlargements he developed from them had not constantly been in front of me, this work would certainly not have been undertaken.

In the final draft, I benefited greatly from suggestions from seven attentive reviewers, especially the most severe of them, my son Pascal, who also gave me a useful refresher on the basic techniques of the integral calculus.

Truth emerges more easily from error than from confusion, according to Francis Bacon. The hypotheses and hasty generalizations advanced in these pages are assumed in the hope that the wrong ones will be easily revealed.

Finally, I express my gratitude to the secretaries, readers, and artists for the care they took with the text, diagrams, and photographs and to the Service des Editions of INRA for having accepted and published this book.

Jean-Pierre André

Contents

Part II: The Technique of Microcasting

Introduction

The effort to integrate our understanding of life into models has the advantage of bringing together disciplines which have a common general objective but have been progressively separated by the specialization of their respective tools and their particular lines of attack. To bring together the study of structures and that of functions and to fill in the "black holes" in our reasoning are among the aims of this effort, and undoubtedly a powerful stimulant of research, provided the models are not too hastily substituted for the real.

In the plant kingdom, the function and structure of the conducting system involved in the ascent of sap have been viewed from different angles, by the physiologist in the longitudinal direction of its functioning, and by the anatomist in the transverse direction, which most clearly reveals the arrangement of tissues. M.H. Zimmermann has especially helped to bring together the two points of view. He proposed a physical description of the water circulation in xylem vessels and, at the same time, attempted to give a realistic picture of these long and narrow tubes, deploring that the microscope was too "myopic" to capture them lengthwise. The ascent of sap in large trees remains a subject of astonishment, thought, and controversy even among scientists. The "plumbing" arrangement is still more or less unappreciated. This book attempts to engage our interest in the structure as much as in the function, drawing the reader's attention to certain aspects of the edification and organization of the vascular system. In order to do this, the book offers the reader a means of seeing the vessels "lengthwise" by the artifice of casting.

Plant histology takes advantage of new tools, some refined, such as laser scanning confocal microscopy, others simpler in their principle, such as vessel casting. Vessel casting gives us a concrete, extended, and highly precise image of the vascular organization of an organ, quite unlike what is seen in histological sections. More precisely, the cast embodies the internal volume of vessels, the volume occupied by the sap. It also embodies the volume of intercellular spaces. In sum, it indicates the volume complementary to what is occupied by the very constituents of the tissues, to which the complete reversion of the solid and hollow parts, with all its finest details, gives a strange beauty. The principle of vessel casting has been used occasionally during the past twenty years in the plant kingdom. Presently the technique is practised by a very small number of laboratories, with variations specific to each, and it is certain to make a significant contribution to the study of vascular organization, particularly of Angiosperms, in future.

The first part of the book offers a general survey of the field of application of this new histological technique, illustrated by the description of vascular structures that it has revealed in the primary and secondary xylem of several tens of Angiosperm species, as well as of Gymnosperms and Pteridophytes. We have emphasized the importance of questions posed by the histogenesis of the newly described structures. We have also proposed hypotheses relating to the concept of procambium-cambium continuum and to that of competition for the space between edge-to-edge cambial sheets. In order to do this, the basic concepts of histology and histogenesis of the xylem are first summarized. Abundant illustrations are provided in the form of original diagrams and photographs of casts, mostly under scanning electron microscope. The diagrams are incorporated along with the text, while the photographs are grouped at the end of each of the five sections.

The second part describes the successive steps in the application of the technique of histological casting, as practised by us, with its advantages and its weak points, and, where necessary, the variants practised by other researchers.

Part I
Potential Applications
of Microcasting

Histogenesis, Structure and Function of the Xylem

OVERVIEW

The cells of terrestrial plants are differentiated[1] into tissues, each with differing morphology and function. Differentiation is absent or weak in Thallophytes and is accentuated in the higher evolutionary plant forms to reach its most advanced stage in the Angiosperms (Table 1). With differentiation, the vital functions of the plant have been progressively allocated to different tissues and, consequently, exchanges of matter and signals have been developed between its various parts. The capture of solar energy and the fixing of atmospheric carbon, for example, have become the specific functions of the photosynthetic tissue and the foliar mesophyll, the absorption of water and dissolved minerals from the soil has become that of the young root tissues at the other end of the plant, and the distribution of absorbed and metabolized substances is that of the vascular tissues, the xylem and phloem, in the species endowed with those tissues, the Tracheophytes.

It has been at least three centuries since the circulation of water and matter in the plant was foreseen from simple observations (Grew, 1682; Malpighi, 1686) and about a century and a half since the vascular tissues were located and named (Hartig, 1837, 1854; Naegeli, 1858; Sanio, 1873, 1874). Research has progressed steadily since, and histogenesis[2] and the structure and functions of the xylem and phloem continue to be fascinating aspects of the study of plant anatomy and physiology.

The xylem and phloem, which were named by the botanist Naegeli, are the pathways of circulation of two flows of matter. The greater flow in terms of mass is the hydric flow of raw sap through the xylem, rising from

---

[1]**Differentiation** is a multi-step process that occurs within a cell during which it becomes morphologically and functionally different from the mother cell. This process realizes one of the possible genetic programmes of cell development, often determined by the chemical or physical signals of the cellular environment and the external environment.

[2]**Histogenesis** is the set of processes of division, growth, and differentiation that leads to the formation of a tissue.

Table 1. Classification of the plant kingdom

Plant kingdom

Thallophytes
Plants not forming stem and having little or no differentiated tissue (algae)

Cormophytes
Plants forming a stem and having differentiated tissues

Bryophytes
Plants forming little or no vascular tissue (Mosses-Hepaticae)

Tracheophytes
Plants forming vascular tissues

Branch

Pteridophytes
Plants forming gametophytes (Lycopoda, horsetails, ferns)

Spermatophytes
Plants derived from:
Seeds* Fertilized ovules* (*Cycas, Ginkgo*)

Sub-branch

Gymnosperms
Plants with naked seeds (Conifers, Gnetales)

Angiosperms
Plants with seeds in an ovary

Class

Dicotyledons | Monocotyledons

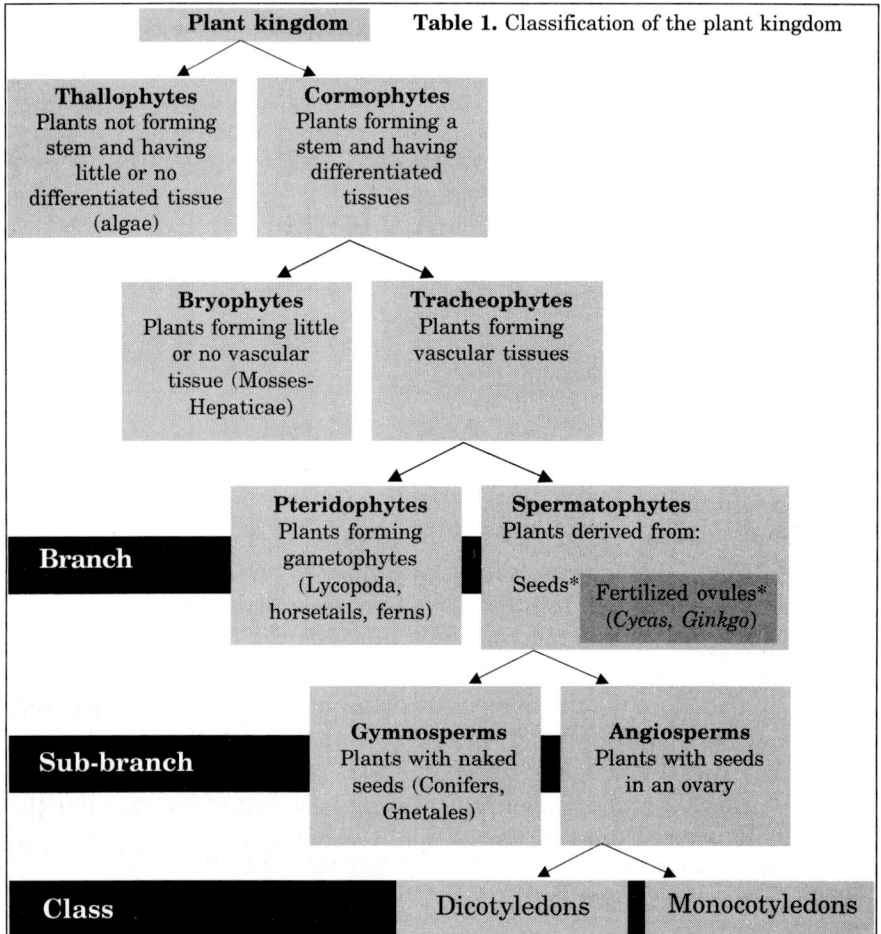

The divisions of the Plant Kingdom are dichotomic at the level of major groups: they are based on the identification of major characters of the anatomy and mode of reproduction of species. Since the appearance of the Cormophytes, or higher plants, the acquisition of these characters over the course of evolution seems to have played a critical role in the extension and spread of plants out of the original aquatic environment.

In the edification of a complex vegetative structure, according to Robert and Catesson (2000), cell differentiation leads to the formation of tissues, the most specialized of which, ensuring the rapid transfer of nutrients and signal molecules from one end of the plant to another, have greatly contributed to the establishment of plants in the terrestrial environment.

The phylum Tracheophytes, or vascular plants, is made up of two groups: (1) Pteridophytes, comprising about 10,000 species, principally the ferns, all survivors of primitive families that are slowly becoming extinct (Emberger, 1960), and (2) Spermatophytes, themselves divided into two sub-groups, the Gymnosperms, a plant group of less than 1000 species, also declining (Emberger, 1960), and the Angiosperms, which constitute the majority of present plants (around 250,000 species). The Tracheophytes present a wide diversity of characters and are the subject of constant phylogenetic reclassification on many points, depending on the importance of new parameters taken into account, particularly genetic ones (Soltis and Soltis, 2000). All these species have the common characteristic of having two kinds of conducting tissue, xylem and phloem. The phylum is named for the resemblance between animal tracheids and the conducting elements of the xylem.

*The seed is a fertilized ovule that maintains exchanges with the mother plant until maturity. A parallel has been established with ovipary and vivipary in the Animal Kingdom (Emberger, 1960).

the soil and supplying all the tissues; most of the water is transpired by leaves and evaporates into the atmosphere. The latter flow, slower and slighter in volume, is the transport of concentrated solutions of carbon metabolites produced by photosynthesis to growing and reserve organs via the phloem.

The two tissues form two closely associated networks, which extend as the plant grows and which draw its architecture with the finest details from the roots to the leaves, flowers, and fruits. Their histogenesis explains their perfect superposition at every point: their cells are in fact derived from divisions of the same meristem, the cambial meristem,[3] which stretches between them like a thin tissue.

In this book, which is devoted to certain aspects of the morphology and spatial organization of the xylem, the presence and role of the phloem are mentioned only incidentally. The overall concepts of histogenesis and histology of the xylem are reviewed in this section so that they can be used to relate the activity of the cambial meristem to xylem structures described in the following sections.

HISTOGENESIS, STRUCTURE AND FUNCTION OF THE XYLEM

The histogenesis of conducting tissues and of the xylem in particular is a difficult process to capture in its spatio-temporal dynamics. To give a brief description, at the tissue scale to start with, let us follow the growth of a shoot of a ligneous Dicotyledon from an apical bud of the preceding year (Fig. 1.1A). From its apex to its lower part, this shoot will comprise at any given moment all the stages of xylem formation.

• Primary and secondary xylem of dicotyledons

At stage 1 of its growth, all the cells of the shoot grow in length. The procambial meristem, partly preformed in the bud, extends the interconnected network of its strands up to the apex. The primary xylem and phloem are differentiated from these strands, the xylem from their inner cells and the phloem from their outer cells (Fig. 1.1, section S1). The bundles, formed of these three tissues, are extended into the leaf petioles.

At stage 2, most of the basal cells of the shoot have *completed* their elongation, while the shoot continues to lengthen by division and growth

---

[3]A **meristem** is a group of cells or a tissue in which the cells divide by mitosis and transmit this property to some of their descendants over a certain number of generations, small in annual species, large in perennial species. The rest of their descendants or derivatives are differentiated into tissues: these tissues are the xylem and phloem in the case of the cambial meristem.

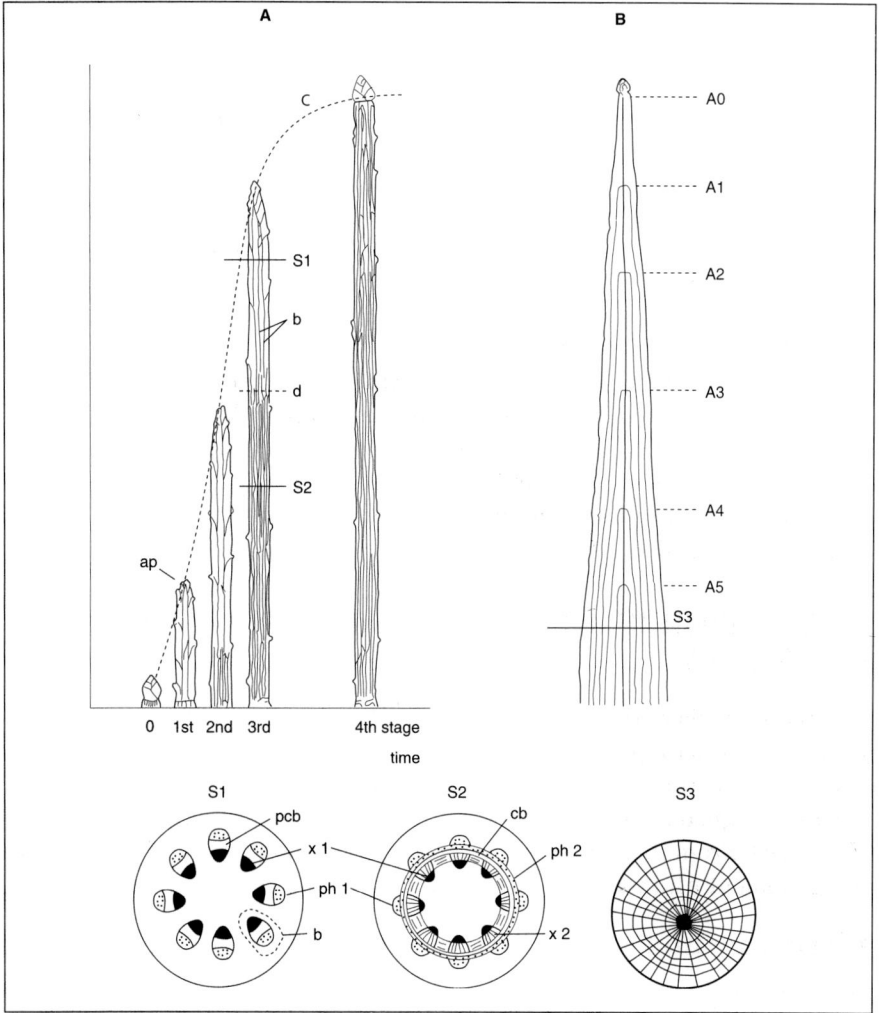

Fig. 1.1. Shoot growth in a ligneous Dicotyledon. A. Successive growth stages between the apical bud of the branch (stage 0) and the end of growth (stage 4). (On the diagrams, the leaves are sectioned at the base of the petioles. The shoot is shown with bark stripped.) The longitudinal growth (curve C) results from the cumulative elongation of all the cells after their formation by division close to the apex (ap). The spaced-out vertical lines at stages 1, 2, and 3 represent the primary bundles (b) in the growing part of the shoot; the horizontal line d delimits the basal part of stage 3, in which the growth is completed: the primary xylem (x1) is progressively ensheathed from the base to the apex by the secondary xylem (x2), represented by short lines close together. The lower part of the figure shows two cross-sections (S1 and S2) respectively above and below the limit d of the two parts: only the meristematic continuum (procambium pcb + cambium cb) and the primary and secondary xylem (x) and phloem (ph) are illustrated. B. A branch is composed of units of successive growth. Seen in axial section, each annual deposit of wood covers and extends the preceding one; seen in cross-section (S3), it forms a supplementary ring (A0, A1, etc.: years of growth).

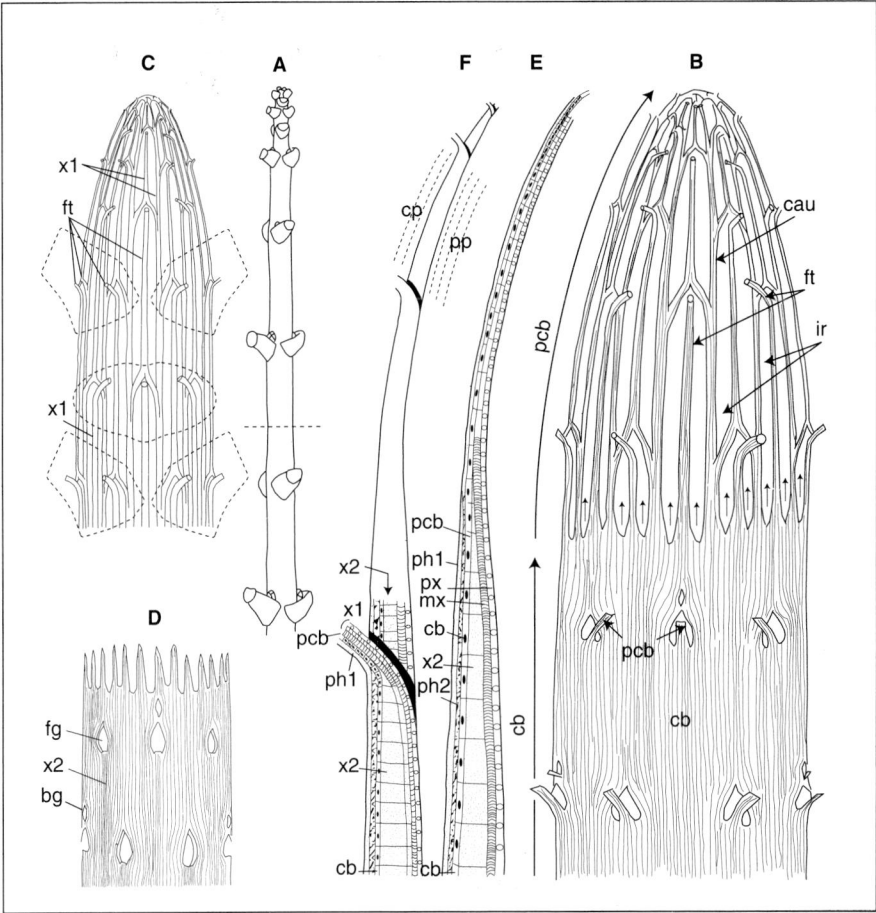

Fig. 1.2. Arrangement of the meristematic procambium-cambium (pcb-cb) continuum and the primary and secondary xylem (x1, x2) in a maple shoot; A. Diagram of the shoot in which the leaves are sectioned. The dottted line marks the present lower limit of the part that is still growing and thus the upper limit of the cambium and wood. In diagrams B, C, and D, the diameter-length ratio of the shoot is dilated by a factor of 8. B. Procambium-cambium continuum: the procambial network produces cauline bundles (cau) the branches of twhich form foliar traces (ft). Lower down, the interfascicular regions (ir) are progressively occupied by the extension of the cambium (arrows); C. The network of the primary xylem (x1), copy of the pcb network, extends from the base to the apex of the shoot; D. The wood or secondary xylem (x2), produced by the cb, progressively ensheaths x1, except at the points of emergence of the primary xylem of leaves (and of non-represented buds) at the "gaps" (foliar gap and bud gap, fg and bg); E. Diagram, in radial section along a cauline bundle, of the superposition of meristems (pcb, cb), primary and secondary phloem (ph1, ph2), primary xylem (x1), itself made up of protoxylem, then metaxylem (px, mx), and finally secondary xylem (x2); F. Analogous section along foliar traces. The whole is included between the pith parenchyma (pp) and cortical parenchyma (cp). The cell dimensions are not related to those of organs.

of the younger cells. The meristem is now composed of a basal part, the *cambium*, and an elongating distal part formed of strands of the *procambium*. While the latter continue to produce primary xylem and phloem, the cambium laterally extends into the *interfascicular regions* and forms a double sheath of *secondary xylem* and *phloem*. After the elongation, the shoot grows in diameter (Fig. 1.1, section S2). The secondary tissues extend towards the apex (stage 3) until they reach the base of the new bud, when the growth of the entire organ is completed (stage 4).

Stage 2 is illustrated in greater detail by the example of a shoot in which the bundles are organized as in the maple (*Acer pseudoplatanus*) (Catesson, 1964). Each node bears two opposite leaves in a decussate arrangement (at right angles) from one node to another (Fig. 1.2A). Each leaf is supplied by three conducting bundles, each bundle having been formed by the union of two branches of lateral bundles (Fig. 1.2B, C). A bundle located partly in the stem and partly in the petiole is called a *foliar trace* (ft). The primary bundles (pcb + x1 + ph1) constituted in a network are separated by interfascicular regions, which are occupied progressively by the cambial meristem. The secondary conducting tissues formed are arranged as two cylindrical sheaths between the primary tissues. Only the latter are extended into the petioles, at least in maple (Fig. 1.2F).

Initiated in the embryo and extending to all the above-ground and underground organs, the *cambial continuum* plays an essential *structuring role* in the plant by producing these two conducting tissues (Fig. 1.3). Moreover, it is involved in the transport of *hormone signals*. The production of primary conducting tissue by the procambium is common to all species; the production of secondary conducting tissue by the cambium is really important only in the arborescent species (Gymnosperms, woody Dicotyledons, some Monocotyledons, and, for the record, quite a number of arborescent fossil Pteridophytes).

The appearance of the cambium, after that of lignin, during the Upper Devonian allowed the development of arborescent forms (Lepidodendrons, *Calamites, Sigillaria*), according to Robert and Catesson (2000). At present, the cambium makes possible the existence of trees, coniferous and broad-leaved. It should be noted, however, that some species without cambium, such as certain bamboos, can reach heights of 15–20 m, while other species that have cambium are naturally trailing plants, such as tomato.

• Differentiation of conducting elements of xylem

The differentiation of a cell derived from divisions of the cambial continuum into a conducting element is of short duration, some hours to a few tens of hours when the spring tissue forms in Angiosperms. It begins with a growth phase, principally in length in the primary xylem and in diameter in the

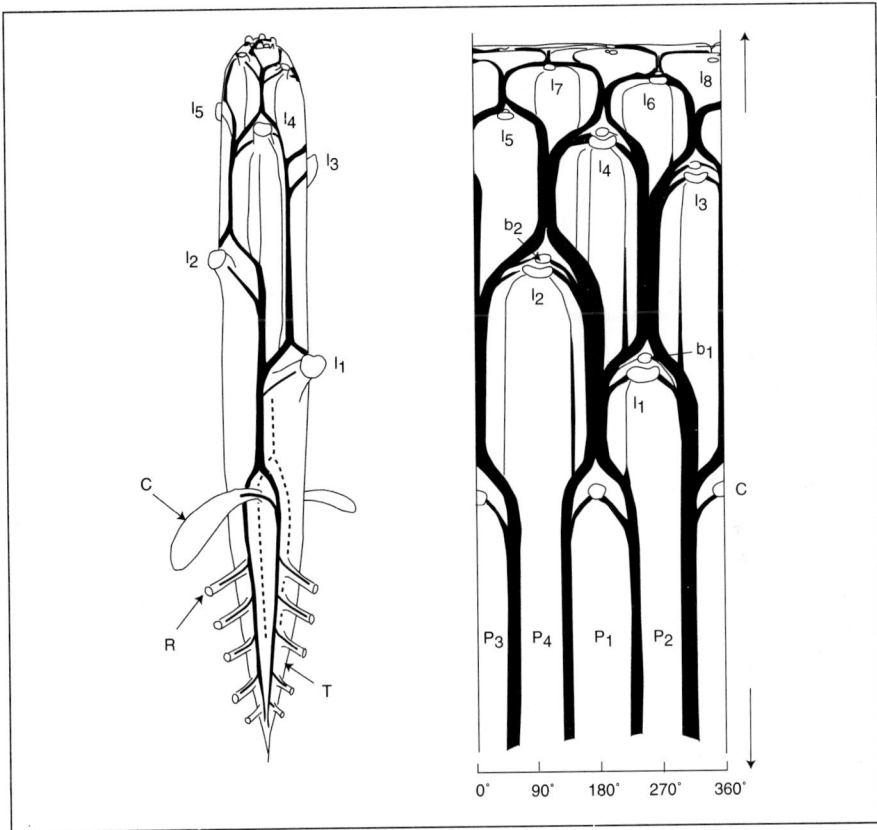

Fig. 1.3. Arrangement of xylem bundles in a tomato seedling. Left, diagram of the seedling, the leaves of which are sectioned (the diameter-length ratio is multiplied by 4). The phyllo-taxy is alternate. (T, taproot; R, lateral roots along the xylem poles; C, cotyledons; l_1, l_2, ..., leaves.) Right, "unrolled" representation of the vasculature (the seedling is represented as a cylinder split along a generating file). The leaf insertions (l_1, l_2, l_3, ...) and bud insertions (b_1, b_2, ...) are indicated. Note the continuity between the vasculature of the taproot (four xylem poles P1 to P4) and that of the stem.

secondary xylem. The second major step is the deposition, in successive layers, of a thick and lignified secondary wall[4] on the inner side of the primary wall. The differentiation ends after the lysis of the cytoplasm and the plasma membrane, i.e., through *cell death*. At the end of the differen tiation process, the only parts of the cell remaining are its rigid and partly perforated secondary wall and the adjacent primary wall, the uncovered

---

[4]**The adjectives** "primary" and "secondary", applied to the xylem and the wall, should no cause confusion. The elongating xylem and the walls of the elongating cell are primary. Th xylem formed after the organ stops elongating and the wall formed after the cell stop elongating are secondary. The cells of x1 and x2 form a primary wall, then a secondary wal

parts of which are thin and pervious to aqueous solutions. This rigid structure, hollow and porous, joined end to end with those of other dead cells, constitutes the conducting system of the xylem. The conducting xylem is the only tissue that acquires its function at the cost of programmed cell death.

The differentiated cells close to the apex form the *protoxylem*. They are differentiated completely *in the course* of elongation. Their secondary wall is composed of rigid rings or whorls of a helix that allow them to stretch lengthwise (Figs. 1.2E and 1.4A). When the cells die and become functional, they undergo first a passive stretching, by the force of elongating adjacent living cells, and then a progressive crushing, when the regions between the rings and helixes no longer present sufficient resistance to lateral thrust. This leads to the "total sacrifice" of the protoxylem, the function of which is taken over by the *metaxylem*.

The future cells of the *metaxylem* begin to differentiate close to the apex but complete their longitudinal growth before forming their secondary wall; at the time they complete their elongation, they are a few millimetres short of the apex (Fig. 1.2E). The metaxylem is the principal and final xylem of mature organs of Monocotyledons. In Dicotyledons, it is a transitional xylem, functional for a few months. In the metaxylem, the secondary wall takes varied forms categorized into a few characteristic types (Fig. 1.4A, 3 to 8, and Pls. 1.6 and 1.7). In all the cases, the secondary wall leaves uncovered more or less extended areas of the primary wall, called pits, through which the sap circulates laterally (Fig. 1.4B, C, E, and Plate 1.8). Moreover, the terminal double primary wall separating two successive elements of the same cell file is generally *perforated* in the metaxylem, more frequently than in the protoxylem (Fig. 1.4D, E). A file of elements, all perforated, constitutes a *vessel*. The non-perforated conducting elements are called *tracheids*.

The conducting elements of the *secondary xylem* or *wood* are then differentiated, without undergoing notable elongation. These are principally the vessel elements (perforated elements) in the Dicotyledons and exclusively tracheids in the conifers (Fig. 1.5 and Plate 1.9B, C). However, their growth in diameter is significant and can in some species greatly exceed their length (Plate 1.9A). The *wood* is deposited in distinct annual rings, made up of a spring part and a summer part in temperate and cold zones, and in more or less continuous concentric layers in tropical species, each new layer or annual ring covering and overlapping the preceding rings (Fig. 1.1B).

The water rising through the wood vessels of the trunk and branches reaches the shoots and leaves, passing laterally through the network of the primary xylem. The functional continuity of the xylem (and phloem)

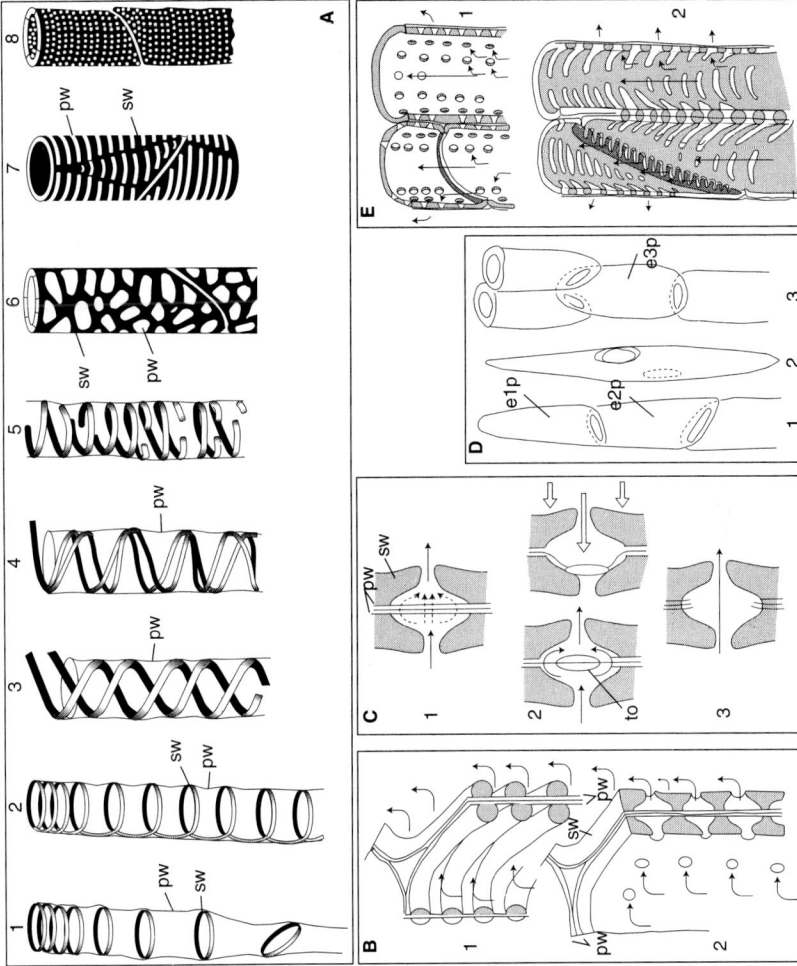

Fig. 1.4. The primary walls (pw) and secondary walls (sw) of conducting elements of the xylem. Pits and perforations. A. Thickening of the secondary walls. (1) Annular thickening of a partly stretched element. Helical thickening: (2) simple helix; (3) double helix; (4) "fissured helix"; (5) interrupted helix. (6) Reticulate thickening. (7) Scalariform thickening. (8) Pitted thickening. B. Circulation of water flows through the primary walls of two adjacent conducting cells between the helixes (1) or in the pits (2) of the secondary walls. C. Water flow through a pair of pits (1), with intact double primary wall (pw) between two vessels (2) pit pair with torus between two tracheids of conifers. The primary wall of the torus margo is highly porous. The torus (to) functions as an obturator when there is a sudden aspiration (large arrows). When the double primary wall (pw) is destroyed (3), the pit functions as a perforation. D. The perforations characterize the elements of the vessel. There are uniperforated elements (e1p) at the tips of the vessels, biperforated elements (e2p) in the body of a vessel, triperforated elements (e3p) at the origin of ramifications: (1) tip of vessel; (2) element with lateral perforations; (3) branched vessel. E. Simple perforations (1) and scalariform perforations (2). The vertical and horizontal flow of the sap is suggested by the arrows.

originates in the functional continuity of the cambial meristem. The *pro-cambium-cambium continuum* concept has been explained clearly by Larson (1982) on the basis of his own observations and those of the major-ity of contemporary researchers.

Moreover, Larson introduced and defined in this continuum the inter-mediate stage of the metacambium, generator of the metaxylem (and metaphloem), while recognizing the artificial nature of such a division and the unclear limits between procambium, metacambium, and cambium. He first confirmed the hypothesis of an *always acropetal*[5] extension of the tips of the procambial strands, prolonged by apical divisions, throughout the process of organ growth.

• Vessels and tracheids

The *vessels* of Angiosperms are highly specialized structures in water con-duction. They are closed ducts of finite length, constituted by the juxtapo-sition of a very large number of dead cells. The water enters and leaves this duct by diffusion through pits in its walls and circulates inside by free convection (Figs. 1.4B, C, D, E and 1.5).

During the formation of a vessel, a long file of several tens, hundreds, or even thousands of cambial derivatives almost simultaneously go through identical processes of differentiation, the result of which is the programmed death of the entire file. However, a single vessel, no matter how long, will

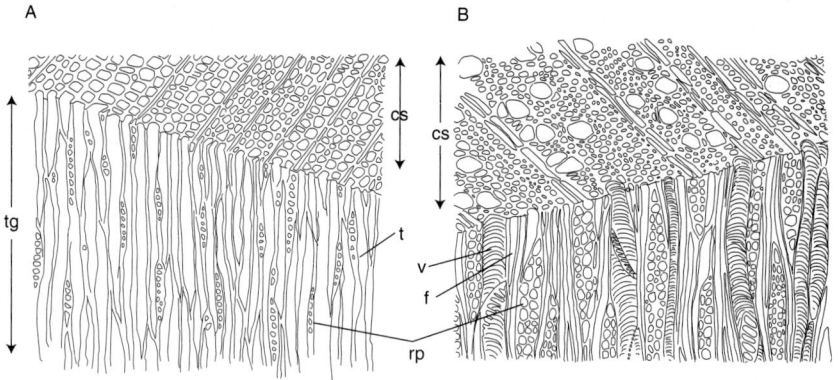

Fig. 1.5. Blocks of conifer wood (A) and broad-leaved wood (B), cut in a cross-section (cs) and a tangential section (tg). A. The wood cells of conifers are tracheids (t) and cells of radial parenchyma (rp) for the most part. B. The wood of broad-leaved trees is made up of fibres (f), vessel elements (v), axial and radial parenchyma and, depending on the species, tracheids and cells intermediate between fibres and tracheids. (Based on photographs by Butterfield and Meylan, 1980.)

---

[5]Acropetal, basipetal: directed respectively towards the tip and towards the base of an organ.

not extend from the roots to the leaves. It is a functional unit included in a vascular whole; that vascular whole extends from one end of the plant to the other. Each newly formed vessel is incorporated in the set that is contemporaneous with it, each annual ring forming its own conducting system. Intensive research is in progress to reveal *in vivo* the inductive role of cellular chemical signals in the differentiation of conducting elements and, at higher levels, in the formation of the functional units i.e., the vessels, and in the edification of an entire vascular system in the plant.

All vessels are not linear files of identical elements. There are a large number of vessels with a very different morphology. The following two chapters are devoted to the description of the variety of these lesser-known forms: one chapter to the primary and the other to the secondary xylem. We will show that these vascular types are not anatomical anomalies but characterize either particular zones of the plant or particular groups of species.

In the wood of conifers, the *tracheids* form a more compact vascular system than the wood vessels of broad-leaved trees, because they constitute the principal cell population of it, the second being the radial parenchyma (Fig. 1.5). The water circulates between adjacent tracheids via their numerous pits with torus (Fig. 1.4C). The principal water flow is established between tracheids of the same generation formed most recently, all of which together constitute a porous, continuous cylindrical sheath, or annual ring, in the roots, trunk, and branches. To some extent, the tracheids function "in parallel", while the elements of the vessels function "in series". In most other respects, however, the performance of the two types of conducting systems is comparable: the sap rises to heights that may exceed 100 m in the vessels of *Eucalyptus regnans* (Dicotyledons) and in the tracheids of *Sequoiadendron giganteum* (conifer).

It should be added that tracheids are not found exclusively in the wood of Gymnosperms: the wood of some species of Angiosperms contains tracheids in a certain proportion to the vessels. Finally, the xylem of Pteridophytes is constituted principally of non-perforated conducting elements, at least at the outset of their functional period.

The longitudinal dimensions of the tracheids are perfectly accessible under microscope, unlike the vessels. Investigations in this field are therefore limited to the approaches presented on p. 85: nevertheless, the results suggest interesting applications of casting from the anatomical and physiological point of view.

The list of species studied is presented in Table 2.

Table 2. The Tracheophytes described in this book

Species	Author	Family	Order	Plate
		Spermatophytes		
ANGIOSPERMS				
Dicotyledons				
Magnolia grandiflora	L.	Magnoliaceae	Magnoliales	3.11
Platanus occidentalis	L.	Platanaceae	Hamamelidales	3.12
Quercus pedunculata	Salisb.	Fagaceae	Fagales	3.15
Malus sp.	L.	Rosaceae	Rosales	3.16
Eucalyptus globulus	Lhérit.	Myrtaceae	Myrtales	3.5, 3.18, 3.19
Rosa sp.	Tourn.	Rosaceae	Rosales	1.8, 3.17, 5.3
Clematis vitalba	L.	Ranunculaceae	Ranunculales	3.10
Cucurbita maxima	Duch.	Cucurbitaceae	Violales	1.9, 5.7
Acer pseudoplatanus	L.	Aceraceae	Sapindales	2.6, 2.7
Juglans regia	L.	Juglandaceae	Juglandales	1.8, 2.8, 3.13, 3.14
Vigna radiata	(L.) Wilczek	Fabaceae	Fabales	5.2
Solanum lycopersicum	L.	Solanaceae	Polemoniales	1.6–1.9, 2.9, 3.7, 3.9, 3.13, 3.17, 5.4–5.6
Capsicum annuum	L.	Solanaceae	Polemoniales	2.9
Ipomoea learii	Paxt	Convolvulaceae	Polemoniales	3.20
Viscum album	L.	Loranthaceae	Santalales	3.21–3.23
Monocotyledons				
Phyllostachys aurea	Carr./A.& C. Riv.	Poaceae	Poales	2.10–2.13
Fargesia nitida	(Mitf.) Nakai	Poaceae	Poales	2.12, 2.13
Phyllostachys viridis	Robert Young	Poaceae	Poales	2.17, 2.18
Bambusa tuldoides	Munro	Poaceae	Poales	1.7, 2.10, 2.15, 2.19
Triticum sp.	L.	Poaceae	Poales	2.10, 2.16
Zea mays	L.	Poaceae	Poales	2.14, 3.17
Smilax aspera	L.	Smilacaceae	Liliales	2.20
Agapanthus umbellatus	Lhérit.	Liliaceae	Liliales	2.21
Phoenix dactylifera	L.	Palmae	Arecales	2.22
GYMNOSPERMS				
Ginkgo biloba	L.	Ginkgoaceae	Ginkgoales	4.8
Pinus mugo	Turra	Pinaceae	Pinales	5.1
Taxus baccata	L.	Taxaceae	Taxales	4.7
Cupressus sempervirens	L.	Cupressaceae	Cupressales	4.7
Sequoiadendron giganteum	(Lindl.) Buchh	Taxodiaceae	Cupressales	4.6
		Pteridophytes		
Selaginella sp.	Spring	Selaginellaceae	Selaginellales	4.1, 4.2
Lycopodiella cernua	(L.) Pichi Sermolli	Lycopodiaceae	Lycopodiales	4.3
Equisetum hyemale	L.	Equisetaceae	Equisetales	4.4
Pteridium aquilinum	(L.) Kuhn	Pteridaceae	Filicales	4.5

The names of orders are borrowed from the classification of Stebbins (1974).

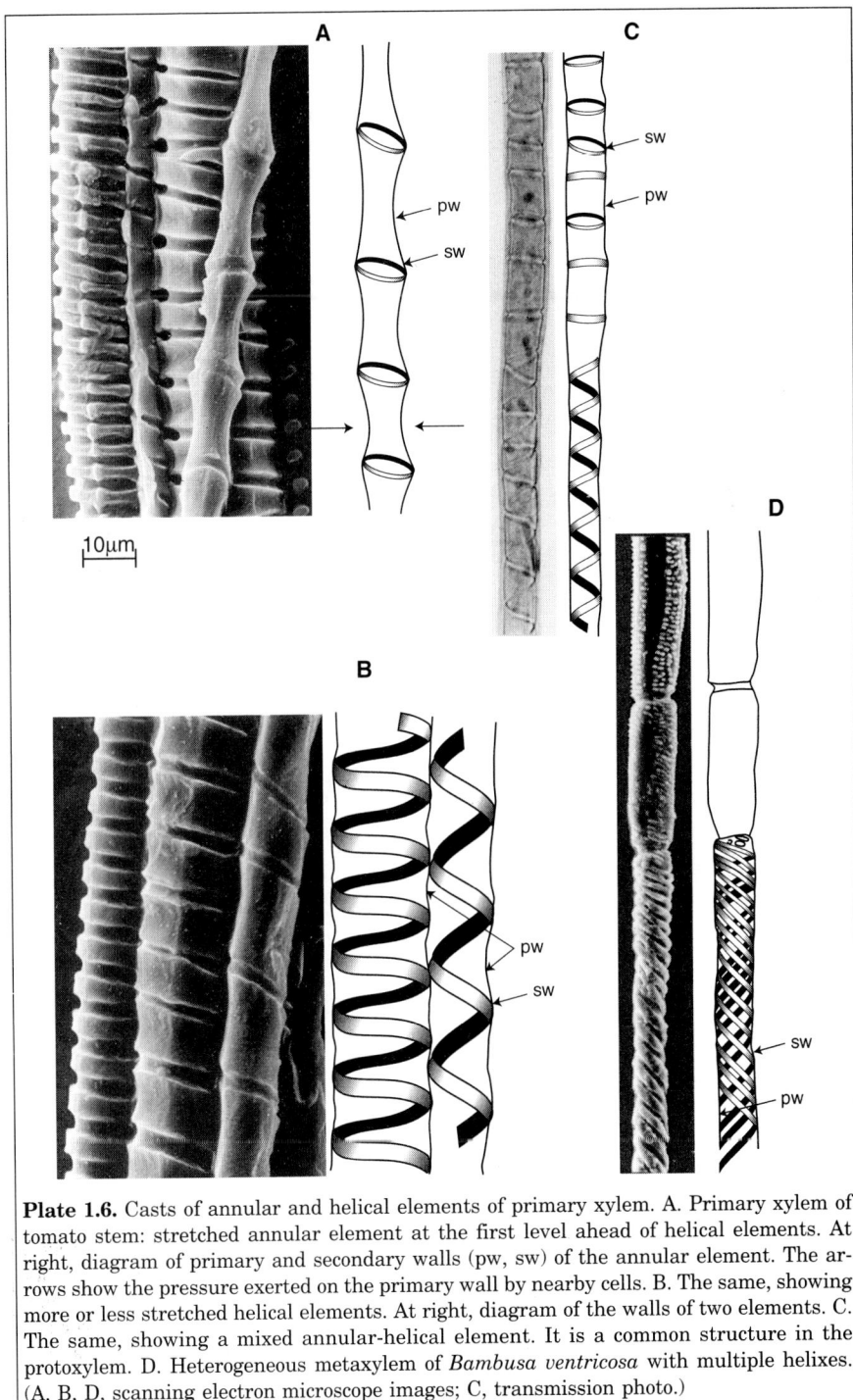

Plate 1.6. Casts of annular and helical elements of primary xylem. A. Primary xylem of tomato stem: stretched annular element at the first level ahead of helical elements. At right, diagram of primary and secondary walls (pw, sw) of the annular element. The arrows show the pressure exerted on the primary wall by nearby cells. B. The same, showing more or less stretched helical elements. At right, diagram of the walls of two elements. C. The same, showing a mixed annular-helical element. It is a common structure in the protoxylem. D. Heterogeneous metaxylem of *Bambusa ventricosa* with multiple helixes. (A, B, D, scanning electron microscope images; C, transmission photo.)

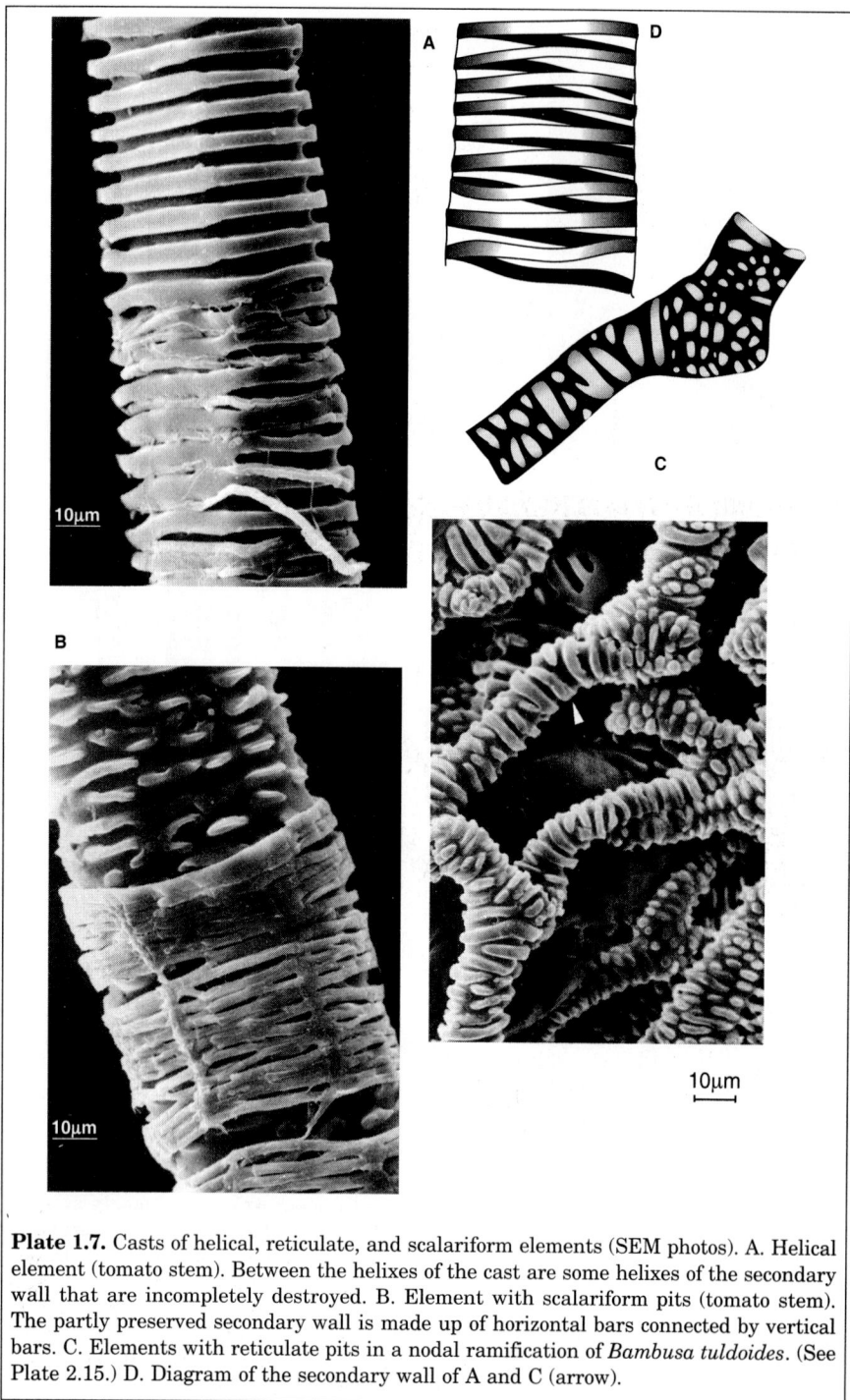

Plate 1.7. Casts of helical, reticulate, and scalariform elements (SEM photos). A. Helical element (tomato stem). Between the helixes of the cast are some helixes of the secondary wall that are incompletely destroyed. B. Element with scalariform pits (tomato stem). The partly preserved secondary wall is made up of horizontal bars connected by vertical bars. C. Elements with reticulate pits in a nodal ramification of *Bambusa tuldoides*. (See Plate 2.15.) D. Diagram of the secondary wall of A and C (arrow).

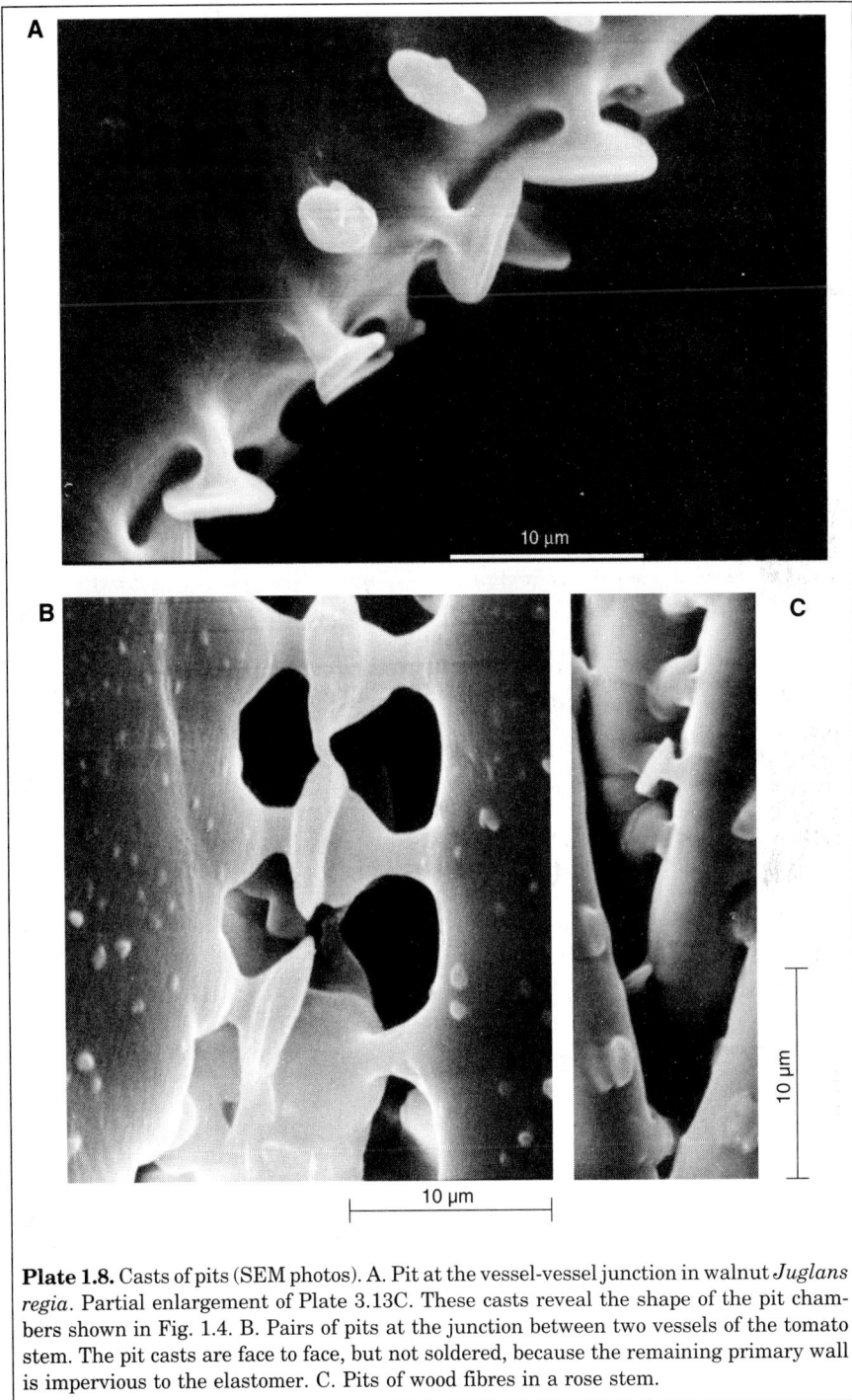

Plate 1.8. Casts of pits (SEM photos). A. Pit at the vessel-vessel junction in walnut *Juglans regia*. Partial enlargement of Plate 3.13C. These casts reveal the shape of the pit chambers shown in Fig. 1.4. B. Pairs of pits at the junction between two vessels of the tomato stem. The pit casts are face to face, but not soldered, because the remaining primary wall is impervious to the elastomer. C. Pits of wood fibres in a rose stem.

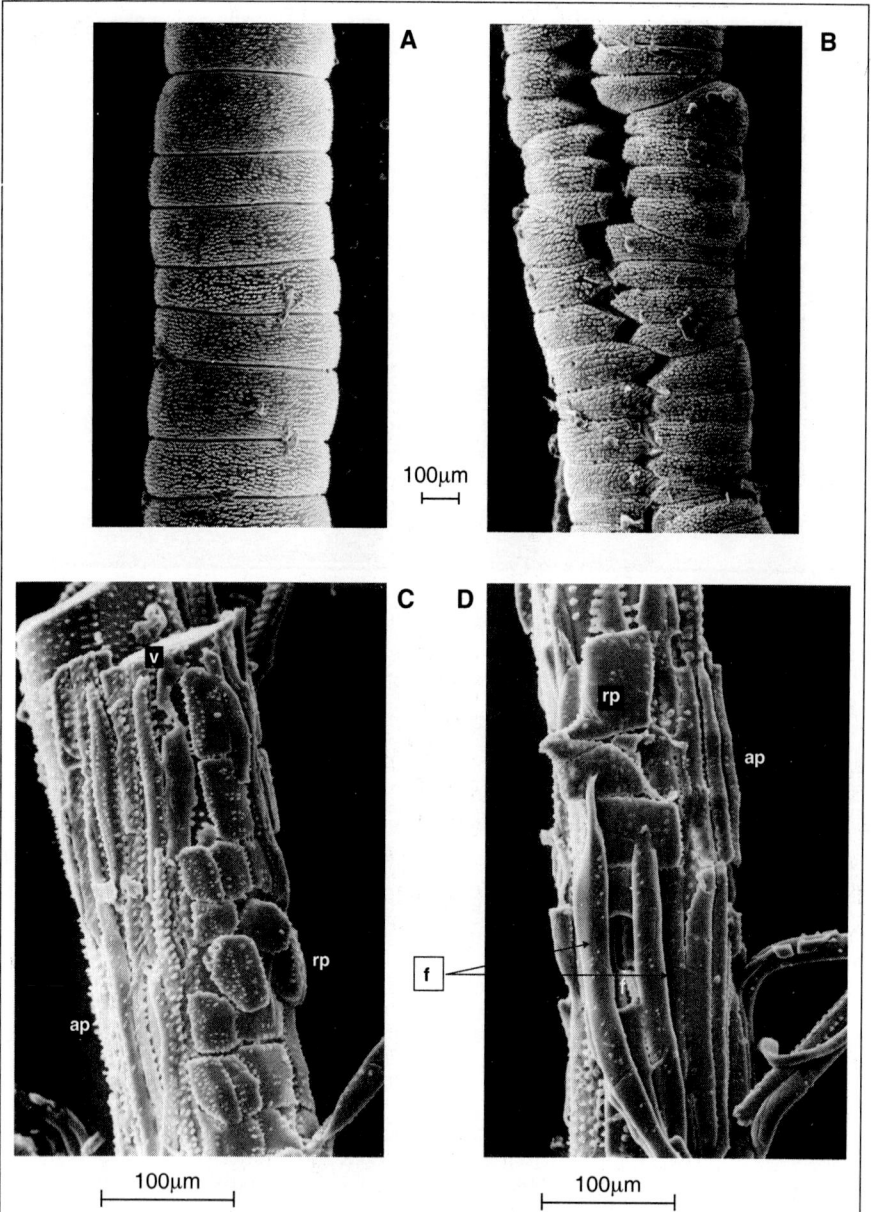

Plate 1.9. A. Cast of secondary vessel elements of squash *Cucurbita maxima*. The diameter of the elements greatly exceeds their height. B. Two segments of adjacent vessels partly separated, showing the mutual zipping of their elements. The pits are densely close and allow lateral water transfers along the entire stem. C. Cast of cells of axial parenchyma (ap) and radial parenchyma (rp) adjacent to the vessels (v) (tomato stem). When the pit walls are highly porous or degraded (Fig. 1.4C), the elastomer stretches from the vessels to penetrate the cells adjacent to it. D. Same cast. Fibres (f), axial and radial parenchyma (SEM photo).

The Primary Xylem of Angiosperms
(Intracellular casts)

ADVANTAGE OF VASCULAR CASTING

The primary xylem of conducting bundles is usually examined and described on the basis of cross-sections, made most often in the internodes. The two vascular structures presented in this section develop in nodal zones that are difficult to access with histological sections. Some authors, whose publications are mentioned in this work, have caught a sight of their complex morphology, but these structures have not been described in the classic treatises on plant anatomy (Esau, 1964; Camefort, 1985; Fahn, 1990). The discovery, location, and precise description of *heterogeneous vessels* and *branched vessels* have been made possible by casting. These results illustrate the interest of the technique. The histogenesis of these two vascular formations is problematic and poses original but highly topical questions. The hypotheses that we advance clearly have a speculative character.

In a more general histological perspective, the studies of Martre (1999) on the vasculature of shoots of *Festuca* was based on the examination of serial sections and vascular casts.

THE HETEROGENEOUS VESSELS

• Definition and description

We have used the terms *transition vessels* (André, 1998) and *heterogeneous vessels* (André et al., 1999) to describe vessels of the metaxylem, all the elements of which do not form the same type of secondary wall. They should more explicitly be called "vessels with heterogeneous segments", since most of their elements are pitted, except over short segments of a few millimetres, in which the elements are reticulate, helical, or even annular. These vessels show thus a local heterogeneity in the differentiation of their elements.

In the Dicotyledons, heterogeneous vessels are observed in the abscission zones, their heterogeneous segment forming exactly in the future plane of abscission. This particular morphology of the metaxylem—the protoxylem and the secondary xylem are homogeneous—seems in fact characteristic of abscission zones of leaves, flowers, fruits, and caducous branches.

In the Poaceae (Monocotyledons), the intercalary growth zones generate curiously the same heterogeneous formations in the metaxylem: in the Bambuseae, for example, the long vessels of the metaxylem are marked by a heterogeneous segment when traversing the base of each internode.

Let us review these various cases.

• Heterogeneous vessels in the abscission zones

Generally, the abscission zones are determined, well before abscission, by local modifications of all the tissues, from the epidermis to the pith parenchyma: reduction of cell growth and of lignification, absence of or late secondary formations. In this section, the conducting elements of the xylem that ensure the vasculature of the caducous organ are described.

At the outset, the study by Böhlmann (1970) of the vascular connection between trunk and branches of conifers and broad-leaved trees should be mentioned. Among his numerous observations, the author mentions briefly and without comment an anatomical detail about the *abscission of young branches* peculiar to pedunculate oak (*Quercus pedunculata*) and to a few other rare broad-leaved trees. He notes that the vessels are pitted in the small branch and in the larger branch that supports it up to the junction. However, within this junction, which contains the future plane of abscission, the vessels are helical in the middle of the zone and scalariform on either side (Fig. 2.1A). The primary or secondary nature of the vessels is not specified. If we acknowledge that these different parts together form continuous files of elements, which seems implied in the description, we have here the characteristic traits of heterogeneous vessels.

Two tropical broad-leaved trees, *Perebea mollis* and *Naucleopsis guyanensis* (Moraceae), also shed their young branches by abscission. In their anatomical description, Koek-Noorman and ter Welle (1976) noted that the vessel elements are more or less isodiametric and scalariform in the plane of the future abscission zone, and cylindrical and pitted above and below this zone. These results also point to the preceding conclusion.

Vascular casts that we have made in the primary and secondary xylem of *flower and leaf abscission* zones present visual proof of the continuity of heterogeneous vessels more conveniently and certainly than do plane sections where the curvature of vessels is three-dimensional.

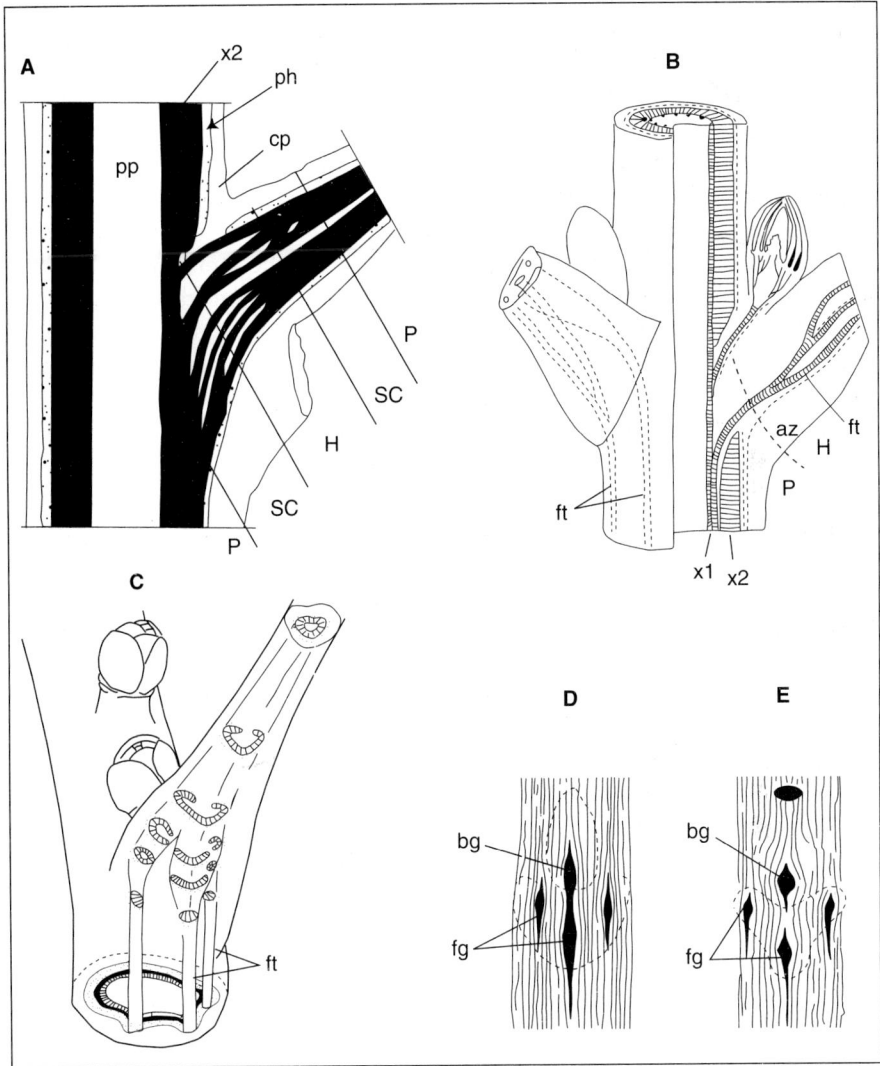

Fig. 2.1. Diagrams of abscission zones of branches and leaves. A. Radial section at the level of insertion of a branch of pedunculate oak (*Q. pedunculata*) two years old (according to Böhlmann, 1970). The xylem and phloem divide into narrow sectors at the base of the branch before forming a continuous cylinder. The vascular regions in which the walls are pitted (P), scalariform (SC), and helical (H) are indicated by the author (x and ph, xylem and phloem; cp and pp, cortical and pith parenchyma). B. Radial section at the level of insertion of a petiole of maple (*Acer pseudoplatanus*). In the symmetrical petiole, two of the three foliar traces (ft) are illustrated. The limit of regions (P) and (H) is indicated, as well as the abscission zone (az) (x1, x2, primary and secondary xylem). C. Perspective view of the union of three foliar traces in the petiole of walnut (*Juglans regia*). D, E. Frontal views of foliar gaps (fg) and bud gaps (bg) respectively of maple and walnut in the wood of the branch.

We have followed, from the stem to the petiole, the foliar traces of the following species with caducous leaves: maple (*Acer pseudoplatanus*) (Plates 2.6 and 2.7), walnut (*Juglans regia*) (Plate 2.8), plane tree (*Platanus acerifolia*), rose (*Rosa* sp.), poke-root (*Phytolacca decandra*), clematis (*Clematis vitalba*), and pubescent oak (*Quercus pubescens*). The secondary xylem present in the cauline part, and sometimes in the petiolar part of the traces, is always interrupted in the zone of future *abscission*. The vascular continuity ensured by the primary xylem is examined by casting: it is composed of continuous homogeneous files of helical elements or continuous heterogeneous files of elements, pitted in the stem, helical in the abscission zone, and helical or pitted in the petiole (André et al., 1999). The degree of heterogeneity varies with the species and even within a single foliar trace (Plates 2.6, 2.7, 2.8). Heterogeneity does not appear in traces of perennial leaves of green oak (*Q. ilex*).

Investigations into *flower and fruit abscission* have been limited to two species of the same family, tomato (*Solanum lycopersicum*) and pepper (*Capsicum annuum*) (Fig. 2.2A, B). The abscission zone studied is located at the junction of the peduncle and the vegetative axis. The development of the primary and secondary xylem has been studied in tomato only, from flowering to an advanced stage of fruiting (André et al., 1999) (Fig. 2.2E). The observations made on serial sections and on casts complement and corroborate those that we have reported for foliar abscission. In the abscission zone, the helical protoxylem is continuous and homogeneous; the metaxylem forms heterogeneous segments (Plate 2.9); the secondary xylem is first interrupted, then extended, and becomes continuous during the enlargement of fruits.

Let us mention for the record the frequently described case of half-annular, half-helical walls, common in the protoxylem, but without any comparable significance in terms of histogenesis (Plate 1.6C).

These descriptions emphasize the morphological traits common to functional vessels of the metaxylem that cross the abscission zones. The observations recorded need to be confirmed by extension to a larger number of species.

• Heterogeneous vessels in the intercalary growth zones

We now turn our attention to the vasculature of Poaceae. The functional xylem of their adult organs, axes and leaves, is made up of the pitted vessels of the metaxylem (see p. 28). The examination of vascular casts reveals that these vessels form a heterogeneous segment in the culms and branches at the base of each internode, and in the leaves, at the junction between the sheath and the blade (Fig. 2.3A, B). In the Poaceae, the growth

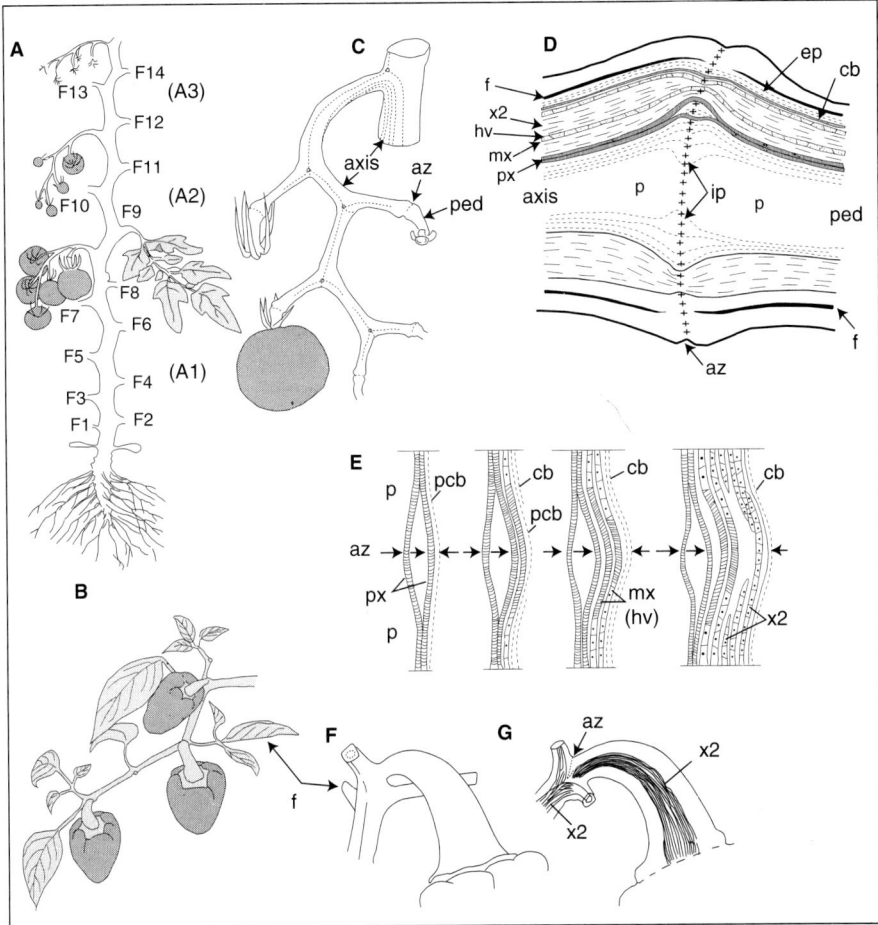

Fig. 2.2. Abscission zone of the fruit pedicel in tomato (A) and pepper (B). Their sympodial stems are composed of successive axes (A1, A2, A3, ... for tomato) axillary to one another. C. In tomato, each stem axis, put out laterally by the following one, ends in a cyme, each branch of which (cyme axis) bears a flower pedicel (ped). D. Axial section in a cyme axis-pedicel junction. The position of heterogeneous vessels (hv) of the metaxylem (mx) between the protoxylem (px) and the wood (x2) is indicated (ep, ip, external and internal phloem; f, pericyclic fibres interrupted in the abscission zone, az). E. Development of the xylem in the abscission zone (az) of the tomato pedicel. From left to right: differentiation of the continuous and homogeneous protoxylem (px), of the continuous and heterogeneous metaxylem (mx), of the homogeneous, secondary xylem (x2), discontinuous at first, then continuous. Late maturation of the procambium (pcb) in the cambial state (cb) in the abscission zone. The pith (p) is at the left of each diagram. F, G. Insertion of the pedicel of pepper and position of the abscission zone.

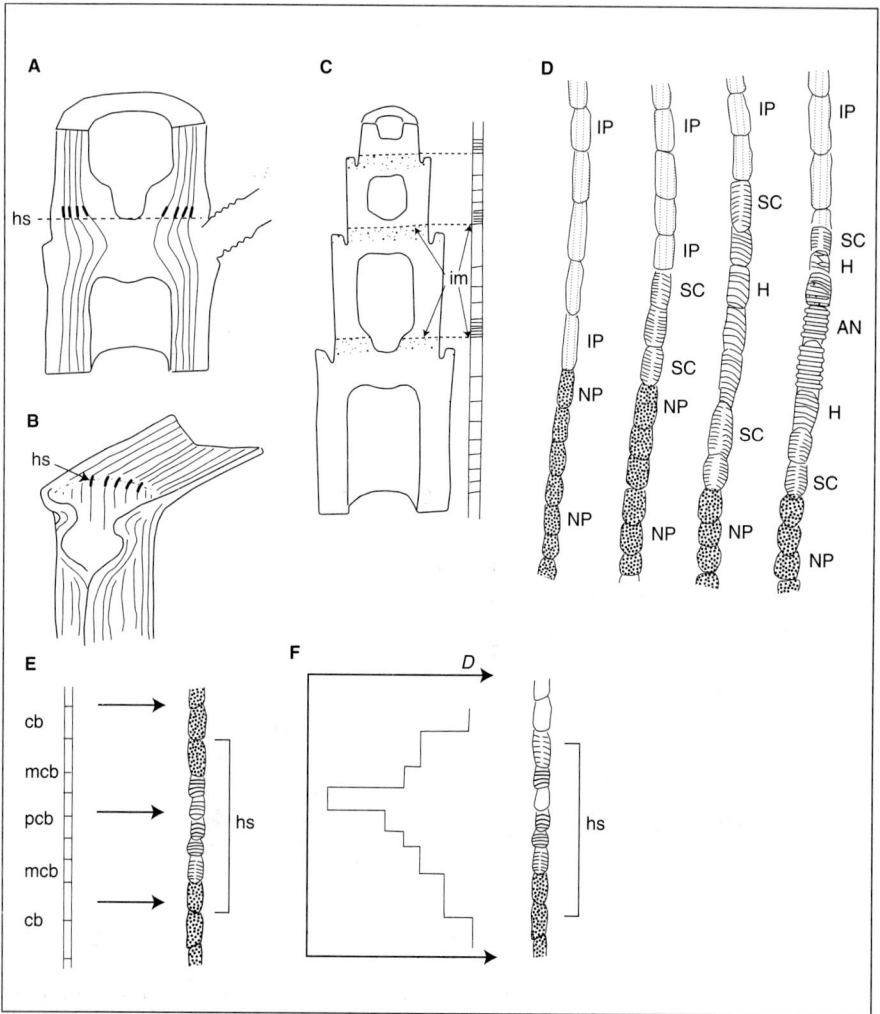

Fig. 2.3. Heterogeneous vessels of intercalary growth zones. A. Axial section in the node of a culm. Location of heterogeneous segments (hs) at the base of the internode. B. Junction between sheath and blade of a Poaceae leaf. Location of heterogeneous segments (hs). C. Axial section in a Poaceae shoot. Location of intercalary meristem zone (im). Next to it is a scale indicating the cell elongation in a vertical line. D. Symmetrical heterogeneous segments of increasing complexity from left to right (NP, IP, nodal and internodal pitting; SC, H, AN, scalariform, helical, and annular wall). E. The meristematic continuum hypothesis suggested by Larson (the heterogeneity of the vessel comes from the heterogeneity of the cambial meristem). F. The differentiation continuum hypothesis suggested by Savidge (the heterogeneity of the vessel comes from differences in the duration D of the expression of genes regulating differentiation).

of culms and leaves results mostly from mitotic activity of intercalary meristems throughout the young organs at the beginning of the growth, then progressively confined to the base of internodes and leaves (Esau, 1965). Therefore, the intercalary meristem extends across the base of internodes and its cells are the last to differentiate and complete their maturation (Fig. 2.3C). The location of heterogeneous segments of the metaxylem in the same transversal zone, at the base of the internodes—a few tens of vessels in wheat, several thousands in a bamboo of 5 cm diameter—led us to see the residual trace of intercalary meristematic activity at that level (André, 1998) (Plates 2.10, 2.11).

The differentiation of elements that constitute a heterogeneous segment has resulted in the juxtaposition of elements with reticulate, scalariform, helical, and annular walls in two more or less symmetrical and complete sequences (Fig. 2.3D). We observe that the limits between two successive wall types do not necessarily coincide with the limits between two elements, because some of them show mixed wall types. In the bamboos, which form several successive axes with numerous internodes, the complexity of the heterogeneous segment increases from the base to the top of the plant: absent in the rhizome and the lowest internodes of culm, the segment presents elements with the largest number of secondary thickening types in the internodes of axillary branches. The differences between species are significant, all other things being equal (Plate 2.10).

The 6000 species of the family Poaceae are divided into subfamilies, the Bambusoideae (bamboos), Festucoideae (most of the temperate herbs and cereals), and Panicoideae (maize, rice, sugarcane, giant reed, etc.). The study of vascular casts has been extended to 25 species of bamboos belonging to 11 different genera (André, 1998), giant reed (*Arundo donax*), and cultivated varieties of wheat, barley, oat, rye, maize, and rice. The anatomical character that we have described has appeared without exception in each species.

In the foliar vasculature, heterogeneous segments form at the base of the blade at the level of the ligule and not at the base of the sheath in the only two species examined, maize and *Phyllostachys aurea*.

• Our hypotheses on the formation of heterogeneous vessels

The question raised by the formation of heterogeneous vessels is one facet of the problem of differentiation of the secondary wall in the xylem in general, more particularly in the wood. We know in fact that the structure of these walls determines the main properties of wood and it has therefore been the object of intensive and ongoing research in several countries (including Canada, Sweden, Japan, the United Kingdom, and France) and speculations on certain steps in the process that are still poorly understood.

In 1996, Savidge suggested that the different types of walls of conducting elements, from the lightest (annular type) to the densest (pitted type), corresponded to successive steps of increasing duration of the expression of genes implicated in the differentiation process (Fig. 2.3F). This hypothesis is challenged by recent observations (Chaffey, 2001; Funada et al., 2001), according to which the determination of a given thickening type seems to be made early and above all be irreversible.

A second hypothesis considers that the heterogeneity of the elements of a vessel results from the heterogeneity of the cell file of the cambial meristem from which it is derived, the former being the image of the latter. According to Larson (1982), the general arrangement of the procambium-metacambium-cambium continuum is oriented from the apex to the base of a growing axis. We suggest that this continuum could take the form of successive symmetrical arrangements such as (... mcb, pcb, mcb...) or (... cb, mcb, pcb, mcb, cb...) each of them being located precisely at the junctions of organs or organ segments that are abscission zones and zones of intercalary growth (Fig. 2.3E) (André et al., 1999).

Remark: This *longitudinal* arrangement of elements with different thickening types in a *single* vessel has nothing in common with the *radial* succession of protoxylem, metaxylem and xylem placed side by side, each vessel showing a different thickening type, as has appeared in anatomy books (cf. Esau, 1965, p. 232).

THE BRANCHED VESSELS

• Definition and description

A branched vessel is made up of several branches linked together by triperforated elements. Its degree of ramification is directly proportionate to the number of triperforated elements (Fig. 2.4A).

In Dicotyledons, the secondary xylem vessels are most often linear; exceptionally they are formed of three branches, two of them being adjacent over their entire length, from a single triperforated element (see p. 55). The primary xylem vessels have not been examined in this regard.

In Monocotyledons, the ramifications are a characteristic trait of the xylem of Poales (single family, Poaceae), but they are not absent in other orders: on a superficial examination, they are found in a few species of Liliales and Arecales.

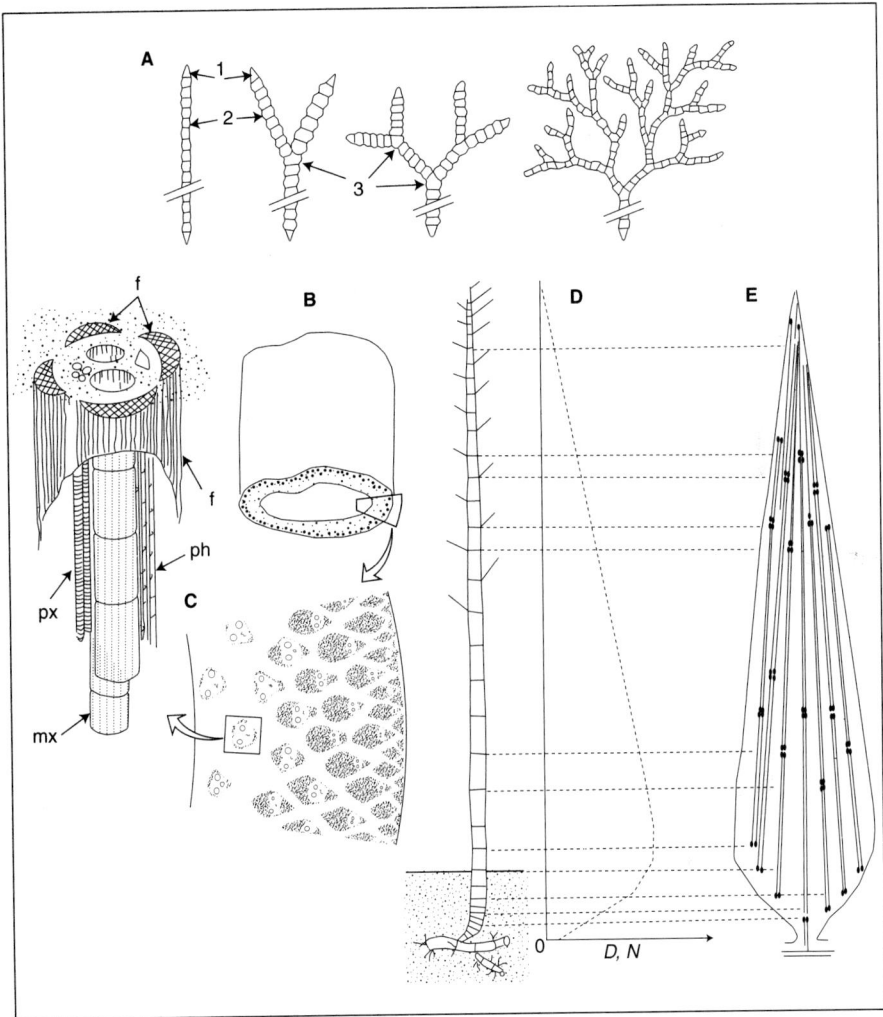

Fig. 2.4. The vasculature of Poaceae. A. Diagrams of linear vessels, simple branched vessels, and more or less regularly dichotomously branched vessels (1, 2, 3, uni-, bi-, and triperforated elements). B. Cross-section of an internode of bamboo, showing the arrangement of conducting bundles. C. In the adult culm, the bundles are composed of the protoxylem (px), which is no longer functional, two vessels of metaxylem (mx), which are functional, and phloem (ph) (proto and meta), surrounded by a fibrous sheath (f). The conducting tissues are better developed and the fibrous sheath less developed in the innermost bundles. D. Concomitant variation of diameter D of the culm and the number N of bundles. E. Diagram of bundles, the culm being represented by a plane surface. The bundles meet end to end in the nodes. The length of bundles is not known.

• The branched vasculature of Poaceae

The description of the vasculature of Poaceae given here is based chiefly on the examination of vascular casts of two species of rhizomatous bamboos, *Phyllostachys aurea* and *Ph. viridis*. Apart from the rhizome-culm junction, studied in *Ph. viridis* only, the vasculature of these two species seems to us to be representative of all the species studied (mentioned on p. 25) (André, 1998).

The vasculature of culms, of entirely primary formation, is organized in bundles, containing, alongside the phloem, a protoxylem and a metaxylem; the former is no longer functional in the adult axes, while the latter is functional. The metaxylem of a bundle is composed of a pair of vessels with pitted elements, the diameters of which are greater (around 100 μm) in the innermost bundles (Fig. 2.4B, C).

The number N of bundles varies with the diameter D of the culm: around one or two hundreds at the junction with the rhizome, it increases in the underground part of the culm up to the soil level, where it reaches a value of about 1500 for a diameter of around 4 cm, then decreases progressively up to the apex (Fig. 2.4D). The bundles are extended over several entire internodes and overlap. Subsequently, each node contains the lower extremities of some bundles and the upper extremities of others, while still other bundles pass over the node, according to proportions that depend on its position (Fig. 2.4E).

Although the organization of bundles, parallel to each other and isolated from each other, may be very simple in the internodes, it is nearly inextricable in the nodes. Each node is the site of xylem junctions, and undoubtedly phloem junctions, between all the bundles that end there. With respect to the xylem alone, Plate 2.11 gives a rough idea of the vascular density of a nodal zone.

Each vessel end is divided in the node into a *ramification* of several tens of branches. Moreover, the vessels crossing the node and ending in other nodes form lateral ramifications there. All the ramifications together constitute a sort of tangled vascular annular plexus at the nodal level (Fig. 2.5 and Plate 2.11).

This annular plexus also contains the ends of branched extremities of vessels of axillary branches. The underground nodes of the culm and of the rhizome also contain the ends of branched extremities of vessels of the root metaxylem (Plate 2.14).

The foliar vasculature, examined in a bamboo (*Bambusa tuldoides*) and in some Festucoideae such as wheat, joins that of the culm by very dense ramifications, in which the protoxylem and metaxylem seem to be mingled (Plates 2.15, 2.16).

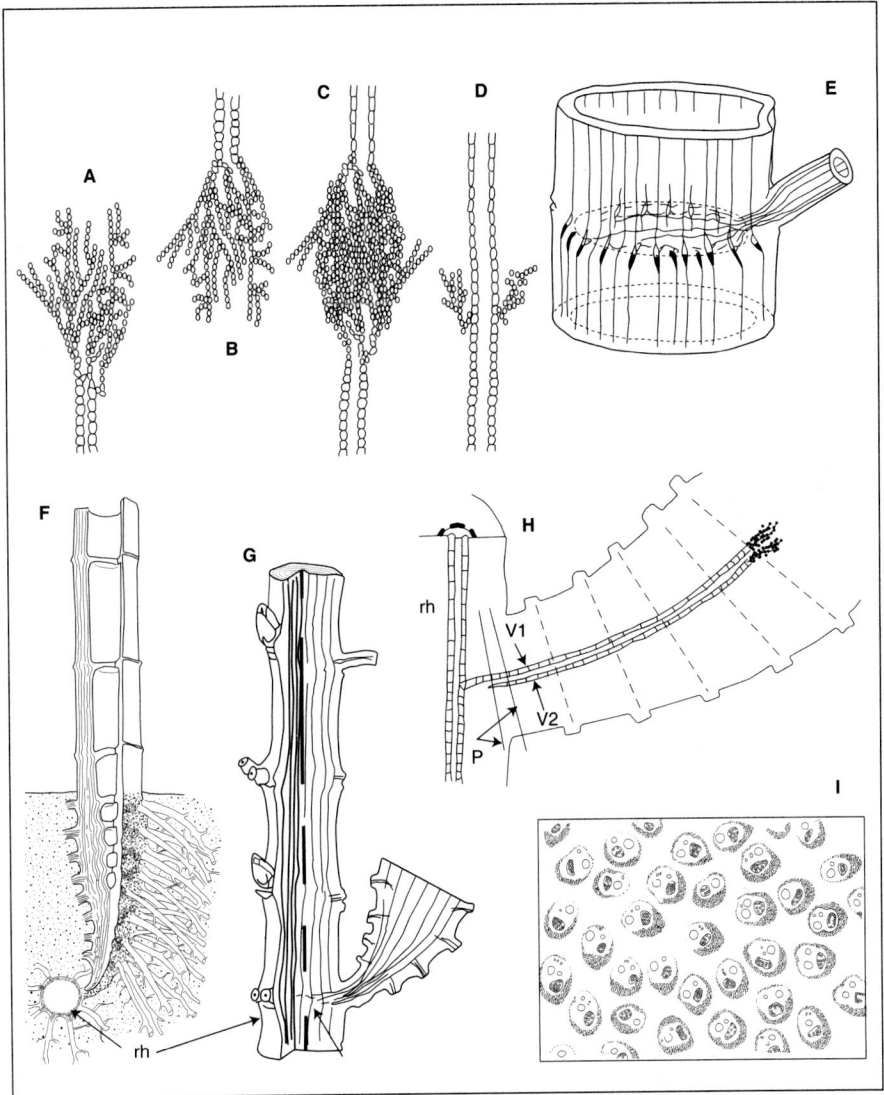

Fig. 2.5. Ramifications in the culm and rhizome-culm junction. A, B. Pairs of ramifications of the metaxylem at the upper and lower extremities of a bundle (free extremities). C. Two pairs of intricate ramifications on two bundles that are continuous end to end. D. Lateral ramifications of a pair of vessels. E. Location of ramifications and vascular ring formed in a node. F. Junction of rhizome (rh) and culm. Rooting of culm at each underground node. G. Partial section of rhizome at the level of the culm junction: the bundles supplying the junction are orthogonal to those of the rhizome (arrow). H. The metaxylem of a bundle at the junction is composed of a lateral branch of a bundle of rhizome (V1) and a free non-connected vessel (V2). V1 and V2 are branched out side by side in a culm node. I. Section in one of the parallel planes (P) of the preceding figure: the bundles seem to contain a single vessel or two vessels of differing diameters.

The technique of selective casting (see Part II, pp. 124–125) suggests that there are no pathways for the casting agent between two distinct ramifications, that is, there are no perforations between intricate ramifications. The innumerable pits of their many elements (Plates 2.12, 2.13) provide these vascular terminations with a considerable exchange surface and hydric transfer. The nodal rings of ramifications are probably zones at which the hydric flow and water potential are balanced between the bundles.

The efficient rooting of the adult culm serves to ensure the self-sufficiency of its water and mineral supply (Fig. 2.5F). Nevertheless, between the culm and the rhizome from which it develops, there are one or two hundreds of conducting bundles (Fig. 2.5G), the principal role of which is certainly phloem transport from the former to the latter. The morphology of the metaxylem vessels in the rhizome-culm junction presents particular features that, because they are few in number, can be examined more easily than in the junctions between the culm and the axillary branches (Fig. 2.5H). The two vessels V_1 and V_2 of a junction bundle are of the same diameter, matched and symmetrical in the underground part of the culm and ending with normal ramifications in a node. In the rhizome, however, they are asymmetrical: V_1 joins by means of a triperforated element to a vessel of the rhizome (itself normally matched in a bundle of this axis), while V_2 ends in a very elongated fusiform extremity that is parallel to the former (Plate 2.17). The longer of the two forms a sort of lateral branch at right angles to a vessel of the rhizome (Plate 2.18), while the shorter is not joined. The asymmetry of the metaxylem pair is evident in the cross-sections of bundles through the diaphragm of the rhizome (Fig. 2.5I).

The nodal vasculature of Gramineae has been the subject of a considerable number of observations based on histological sections, most often cross-sections (Percival, 1921; Bugnon, 1924; Sharman, 1942; Kumazawa, 1961; Patrick, 1972; Zee, 1974; Chonan, 1976; Busby and O'Brien, 1979; Yulong and Liese, 1997). Although the general outlines have been drawn, the diagrams that have been given show many inaccuracies; casting can be used to obtain a more correct interpretation of the sections, at least with respect to the xylem.

• Hypothesis on the formation of ramifications in the Poaceae

In the nodes of growing shoots, examined in *Bambusa tuldoides*, the ramifications of the metaxylem vessels seem, in the casts, less developed than those of the nodes of the adult culm (Plate 2.19). There, the vascular branches are few and short, as if the differentiation of the conducting elements occurs primarily in the vessel to extend progressively into the elements of the ramification during the maturation of the culm. The observation suggests a step-by-step process of induction of cell differentiation, but in an apparently random distribution of branching points, i.e., of triperforated elements. This process may result in the increasing intricacy of ramifications and consequently in the vertical and horizontal circulation of sap through dead conducting cells.

Remark: As the ramifications of the metaxylem develop, other morphologically very different ramifications, which join the culm to the vasculature of active leaves, are already established and functional in the same nodes (Plates 2.15, 2.16). The secondary walls of their elements are helical, reticulate, and pitted. Their complex structure has not yet been elucidated.

• Vascular ramifications in other orders of Monocotyledons

Apart from the order Poales, our research has been limited to a very small number of species in the large orders Liliales (8000 species) and Arecales (4000 species).

Smilax aspera (sarsaparilla) and *Tamus communis* are two climbing Liliales, in which the anatomy of the stem is typical of Monocotyledons (Plate 2.20). In *Smilax*, the adult stem comprises 50 to 100 bundles each containing two similar metaxylem vessels. In the nodes, the vessel pairs close to insertions of axillary branches form pairs of narrow ramifications that are poorly developed, along which the non-ramified extremities of the axillary vessels are closely positioned (Plate 2.20). The metaxylem is not ramified in the second species.

The rhizome of *Agapanthus umbellatus* (Liliaceae), is firmly rooted in the soil by numerous vigorous roots. It produces several tens of distichous alternate leaves over several years before the single inflorescence is formed. When the lower leaves are shed, leaf scars and two rows of axillary buds remain. The foliar vasculature extends into the central cylinder in the form of successive divergent "sheaves": it is composed of tracheids that are resistant to casting. The root vasculature opens out under the cortex, then penetrates deeper into the central cylinder. The metaxylem vessels present a ramification in which only the base can be cast (Plate 2.21). As the example of agapanthus reminds us, vascular casts do not take into account the vasculature constituted by the tracheids.

In the root vasculature of the date palm, *Phoenix dactylifera* (Palmaceae, Arecales), we find ramifications (Plate 2.22) with morphology comparable to that of maize (Plate 2.14). Quite regularly dichotomic, the ramification of the root metaxylem first produces short elements in which the diameter decreases at each new branching, and then very long elements with very small diameters in the more distal parts of the branches.

In the cast of the dense vasculature inside the base of the stem, or plateau, no ramifications appear. Each branch of the root metaxylem appears to join a vessel of the plateau.

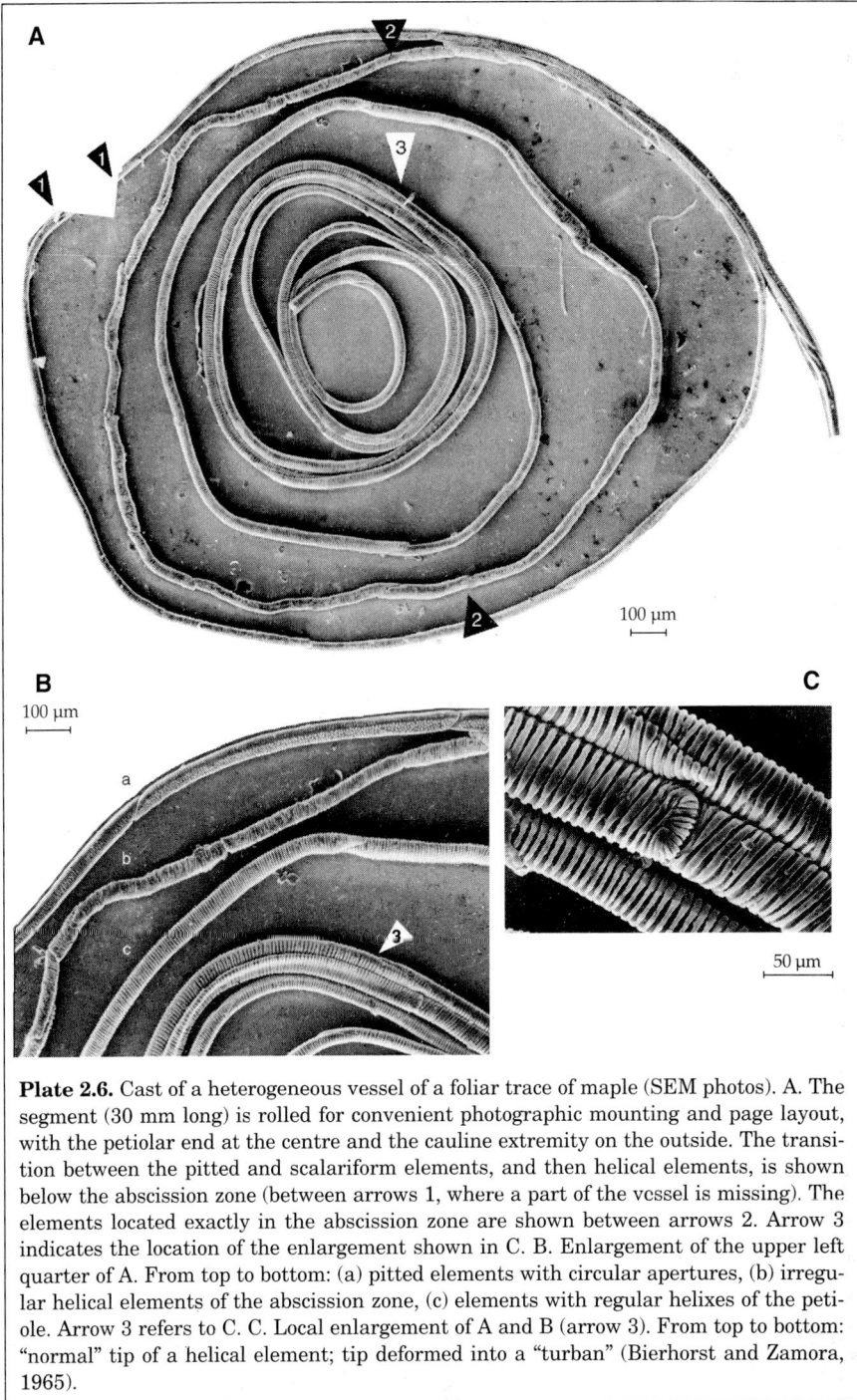

Plate 2.6. Cast of a heterogeneous vessel of a foliar trace of maple (SEM photos). A. The segment (30 mm long) is rolled for convenient photographic mounting and page layout, with the petiolar end at the centre and the cauline extremity on the outside. The transition between the pitted and scalariform elements, and then helical elements, is shown below the abscission zone (between arrows 1, where a part of the vessel is missing). The elements located exactly in the abscission zone are shown between arrows 2. Arrow 3 indicates the location of the enlargement shown in C. B. Enlargement of the upper left quarter of A. From top to bottom: (a) pitted elements with circular apertures, (b) irregular helical elements of the abscission zone, (c) elements with regular helixes of the petiole. Arrow 3 refers to C. C. Local enlargement of A and B (arrow 3). From top to bottom: "normal" tip of a helical element; tip deformed into a "turban" (Bierhorst and Zamora, 1965).

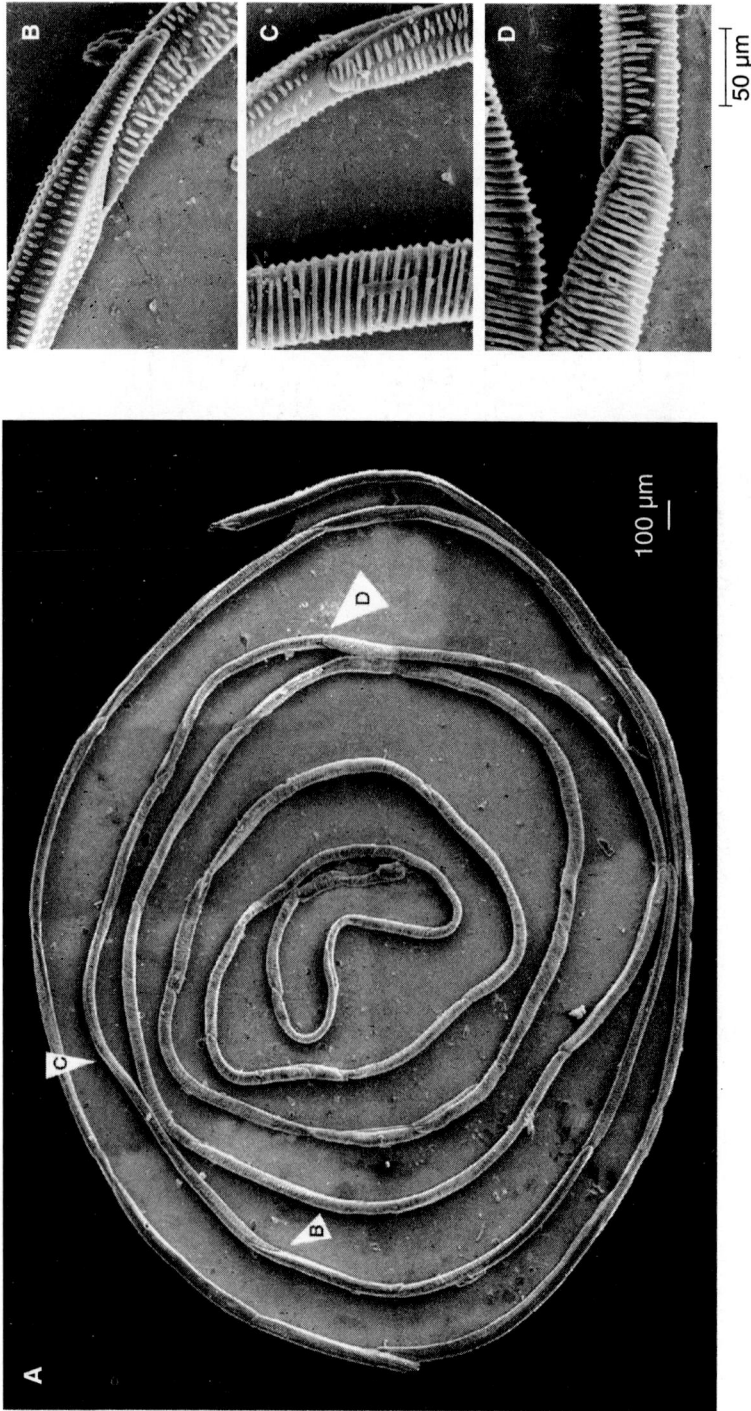

Plate 2.7. Second casting of heterogeneous vessel of foliar trace of maple (SEM photo). A. The segment (27 mm long) differs from the preceding one in its more gradual transition between the pitted and helical elements. Some junctions between successive elements are shown in detail in B, C, D. B. Enlargement of a junction between elements with small and large pits of a scalariform type. C. At right, type of junction similar to B. D. Junction between element of a scalariform type and one of a helical type.

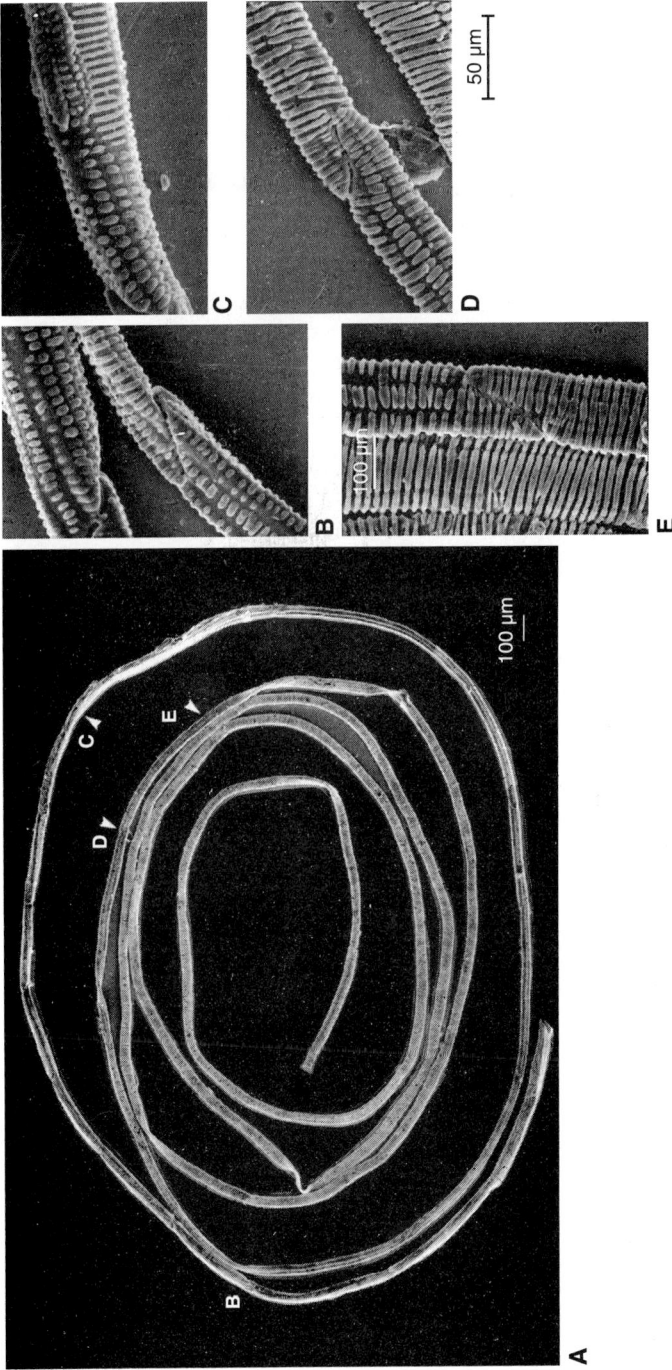

Plate 2.8. Segment of heterogeneous vessel of the foliar trace of walnut (SEM photo). A. The segment (40 mm long) was taken and arranged like those in Plates 2.2 and 2.3. The transition between the pitted elements of the stem and the helical elements of the petiole is gradual. The letters refer to the enlargements. B. Enlargement of a junction between pitted elements. C. Enlargement of a transition between circular and scalariform pitted elements. D. Enlargement of a junction between a scalariform element and a helical one. E. Same as D, and helical elements at right.

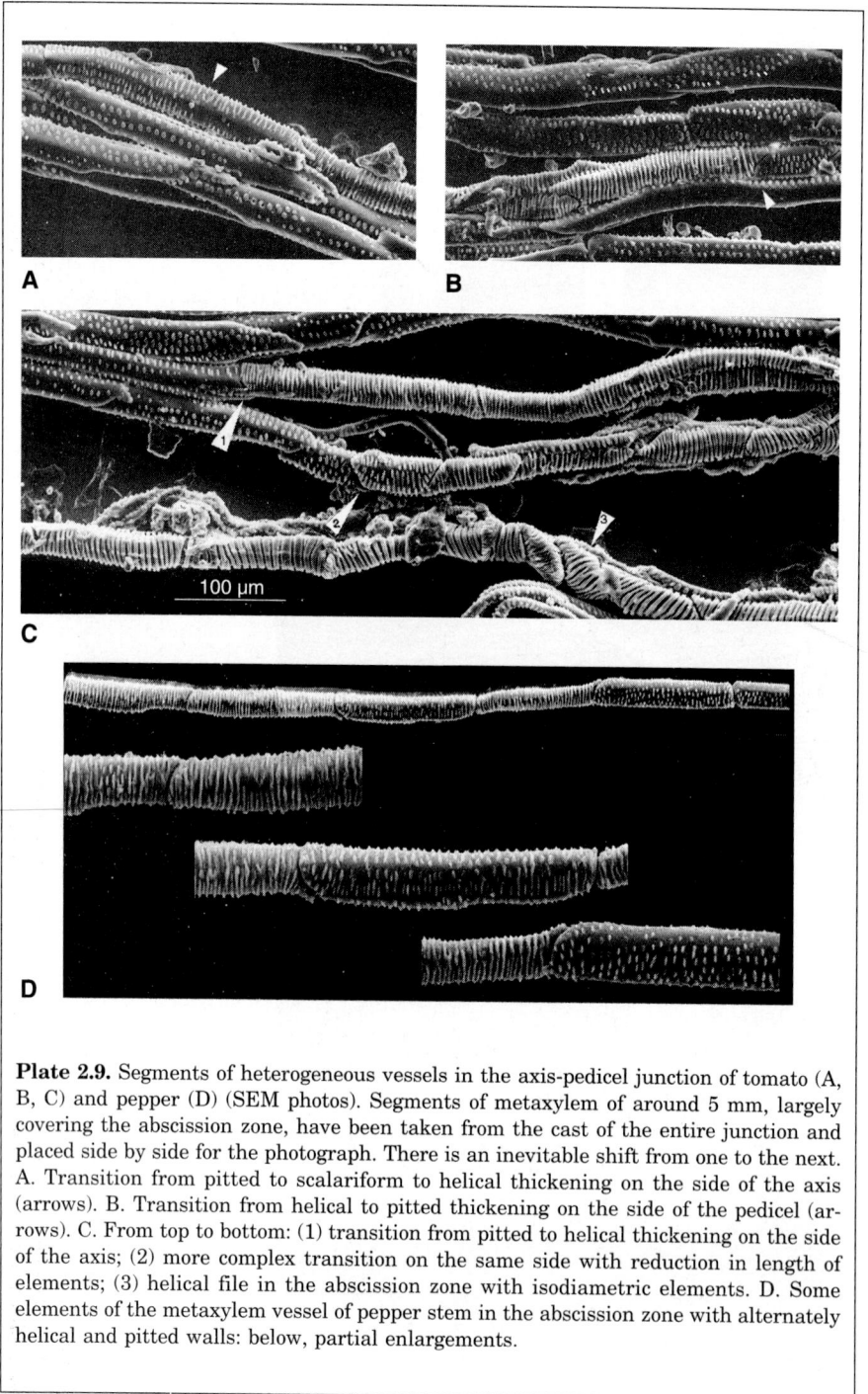

Plate 2.9. Segments of heterogeneous vessels in the axis-pedicel junction of tomato (A, B, C) and pepper (D) (SEM photos). Segments of metaxylem of around 5 mm, largely covering the abscission zone, have been taken from the cast of the entire junction and placed side by side for the photograph. There is an inevitable shift from one to the next. A. Transition from pitted to scalariform to helical thickening on the side of the axis (arrows). B. Transition from helical to pitted thickening on the side of the pedicel (arrows). C. From top to bottom: (1) transition from pitted to helical thickening on the side of the axis; (2) more complex transition on the same side with reduction in length of elements; (3) helical file in the abscission zone with isodiametric elements. D. Some elements of the metaxylem vessel of pepper stem in the abscission zone with alternately helical and pitted walls: below, partial enlargements.

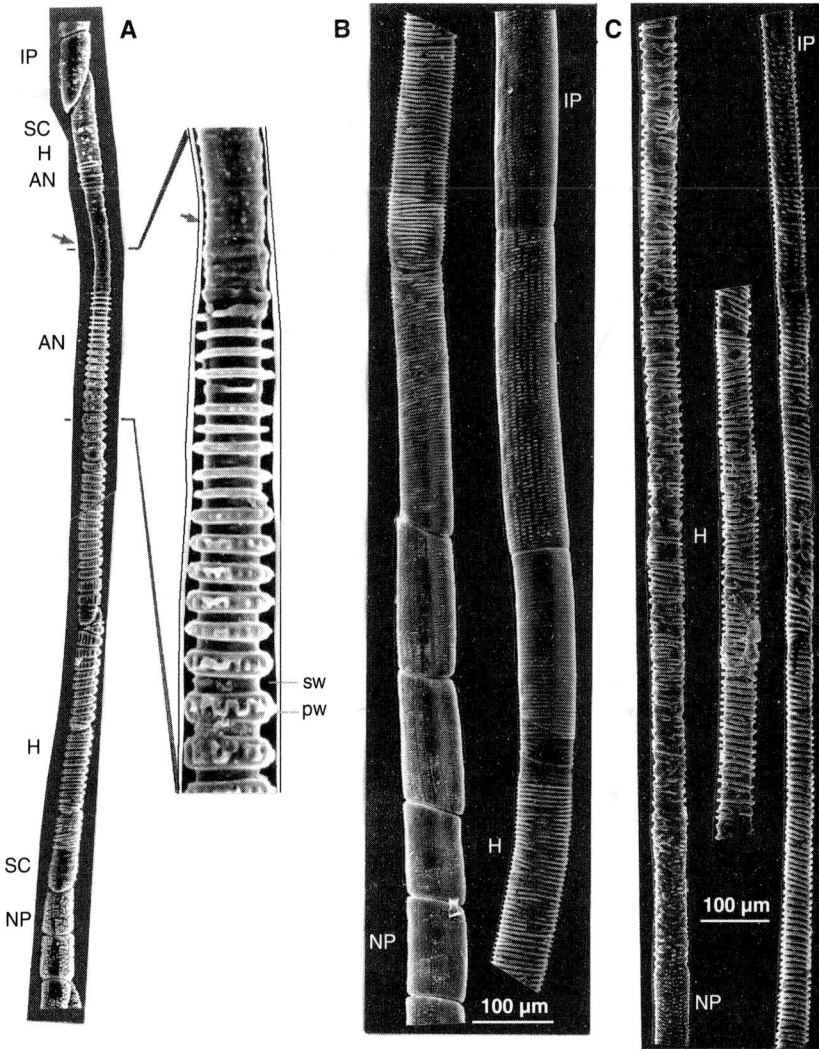

Plate 2.10. Heterogeneous segments in the culm of bamboos (SEM photos). A. *Bambusa tuldoides*. More or less symmetrical segment presenting all the thickening types, sampled from a second-order axillary branch. From bottom to top, nodal elements with circular pits (NP), followed by elements with scalariform pits (SC), helical walls (H), annular walls (AN), absence of secondary wall thickening (arrow), annular, helical, scalariform, and internodal element with circular pits (IP). A portion enlarged: note that the annular thickenings of the secondary wall (sw) are located between the actual annular thickenings of the cast. The relief that appears on the outer edge of the latter is not explained. (pw, primary wall.) B. *Phyllostachys aureosulcata*. From left to right and from bottom to top: NP-H-IP. C. *Triticum* sp. The long heterogeneous segment of wheat is sectioned in three parts, the lower at left, the upper at right, oriented in the direction of the culm, with NP at bottom left, IP at top right. The H segment is very long and the cast partly defective.

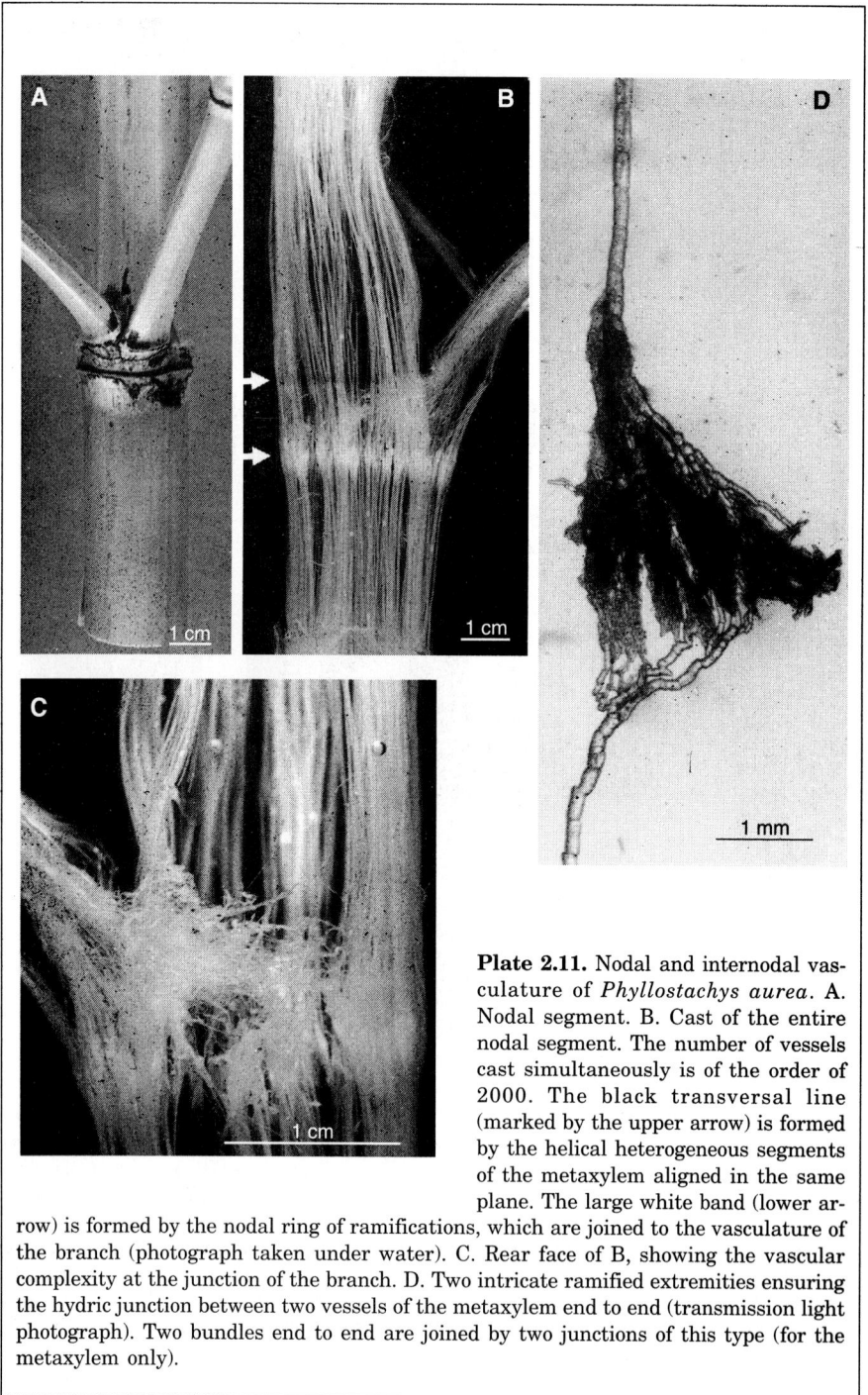

Plate 2.11. Nodal and internodal vasculature of *Phyllostachys aurea*. A. Nodal segment. B. Cast of the entire nodal segment. The number of vessels cast simultaneously is of the order of 2000. The black transversal line (marked by the upper arrow) is formed by the helical heterogeneous segments of the metaxylem aligned in the same plane. The large white band (lower arrow) is formed by the nodal ring of ramifications, which are joined to the vasculature of the branch (photograph taken under water). C. Rear face of B, showing the vascular complexity at the junction of the branch. D. Two intricate ramified extremities ensuring the hydric junction between two vessels of the metaxylem end to end (transmission light photograph). Two bundles end to end are joined by two junctions of this type (for the metaxylem only).

Plate 2.12. Nodal ramifications of the metaxylem vessels of two bamboos: *Phyllostachys aurea* (A, B) and *Fargesia nitida* (C) (SEM photos). A. *In situ*, the ramifications have a more compact arrangement (diagram at right). This is spread out to facilitate the examination of branches. The casting medium penetrates the section of the vessel up to the limits of the ramification (several hundreds of perforated elements). B. Enlargement of the inset in A. The arrow refers to enlargement 2.13A (following plate). C. Partial view of a terminal ramification (around 20% of the whole); in this species, the elements are cylindrical and regularly arranged in the ramification.

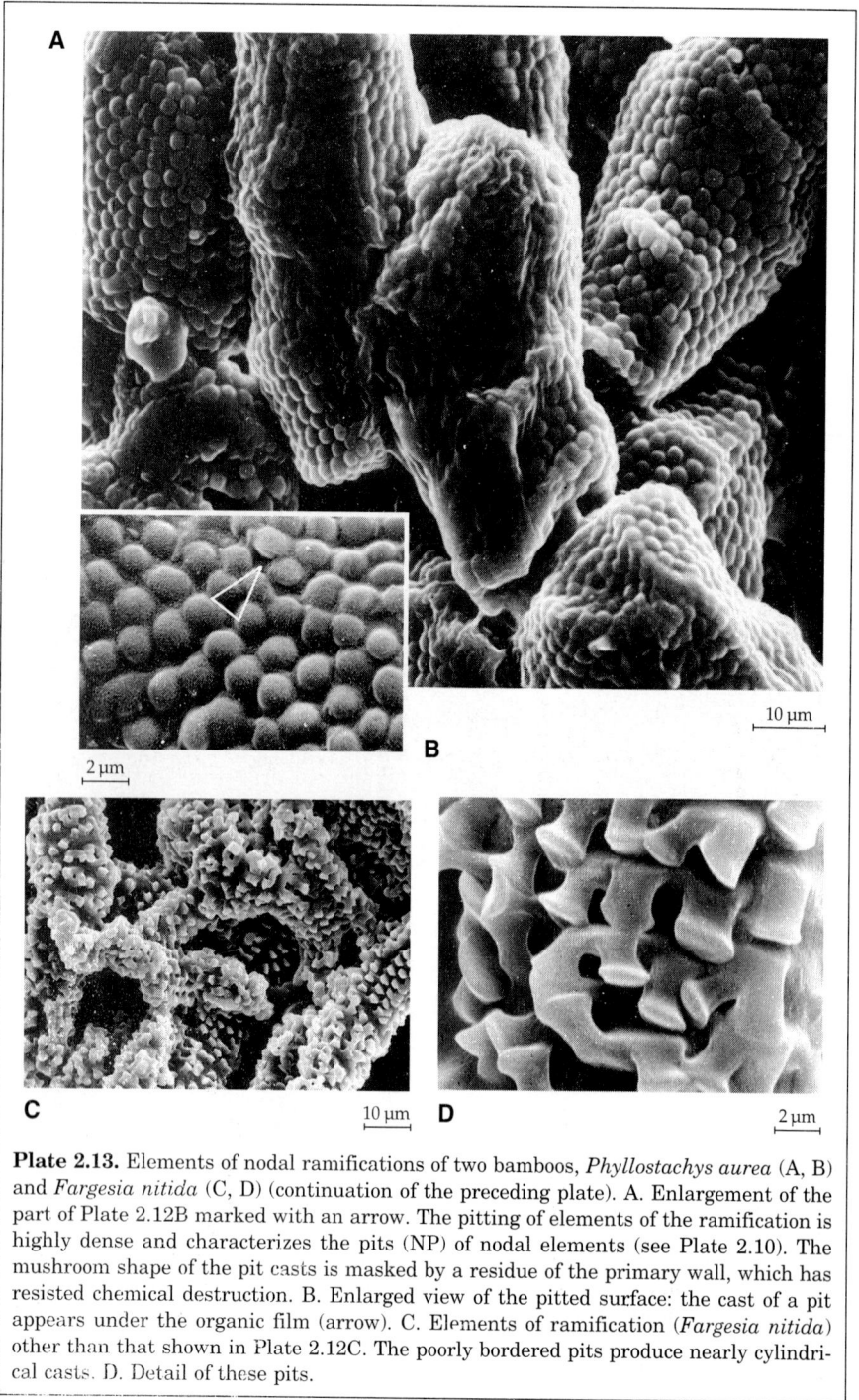

Plate 2.13. Elements of nodal ramifications of two bamboos, *Phyllostachys aurea* (A, B) and *Fargesia nitida* (C, D) (continuation of the preceding plate). A. Enlargement of the part of Plate 2.12B marked with an arrow. The pitting of elements of the ramification is highly dense and characterizes the pits (NP) of nodal elements (see Plate 2.10). The mushroom shape of the pit casts is masked by a residue of the primary wall, which has resisted chemical destruction. B. Enlarged view of the pitted surface: the cast of a pit appears under the organic film (arrow). C. Elements of ramification (*Fargesia nitida*) other than that shown in Plate 2.12C. The poorly bordered pits produce nearly cylindrical casts. D. Detail of these pits.

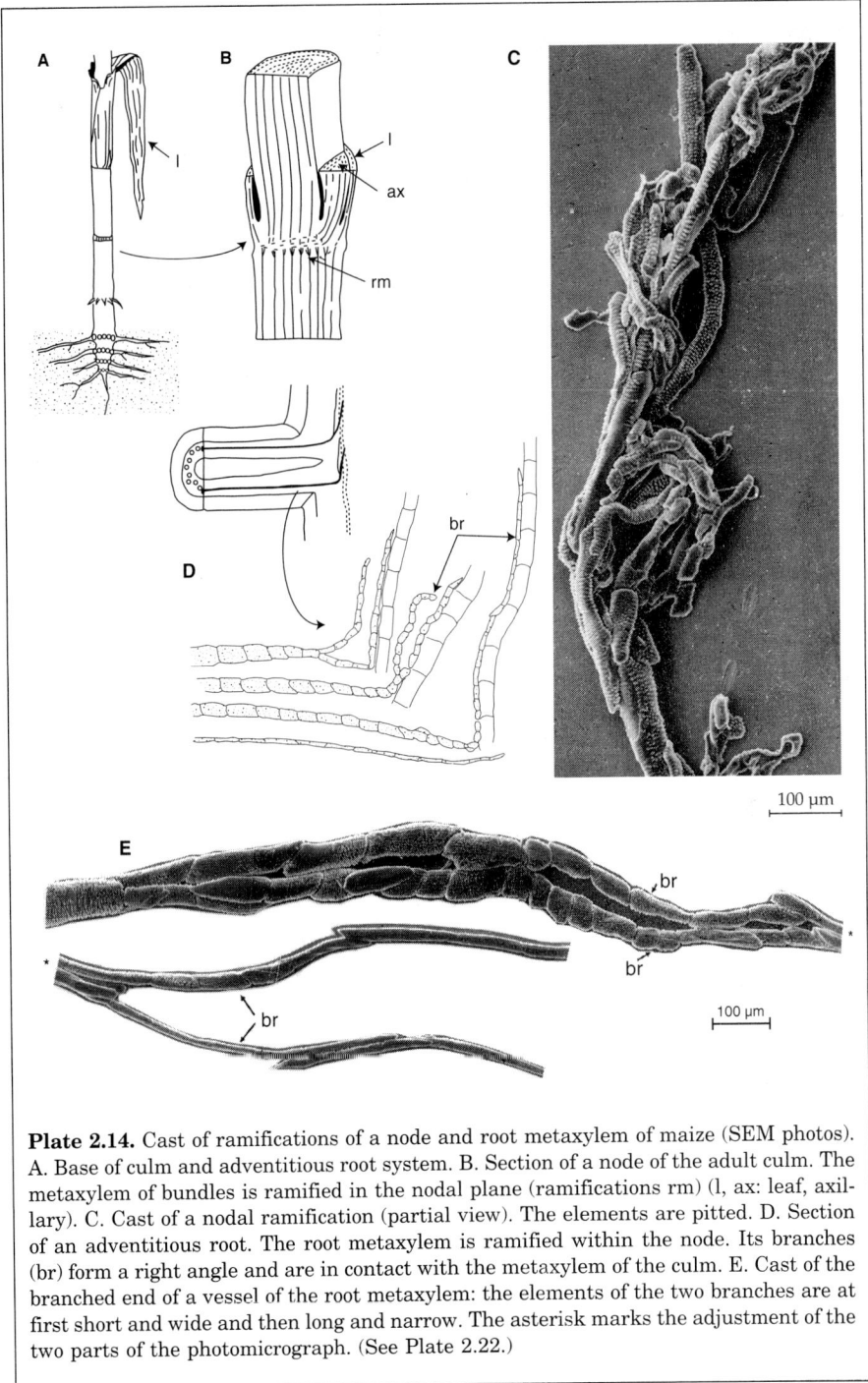

Plate 2.14. Cast of ramifications of a node and root metaxylem of maize (SEM photos). A. Base of culm and adventitious root system. B. Section of a node of the adult culm. The metaxylem of bundles is ramified in the nodal plane (ramifications rm) (l, ax: leaf, axillary). C. Cast of a nodal ramification (partial view). The elements are pitted. D. Section of an adventitious root. The root metaxylem is ramified within the node. Its branches (br) form a right angle and are in contact with the metaxylem of the culm. E. Cast of the branched end of a vessel of the root metaxylem: the elements of the two branches are at first short and wide and then long and narrow. The asterisk marks the adjustment of the two parts of the photomicrograph. (See Plate 2.22.)

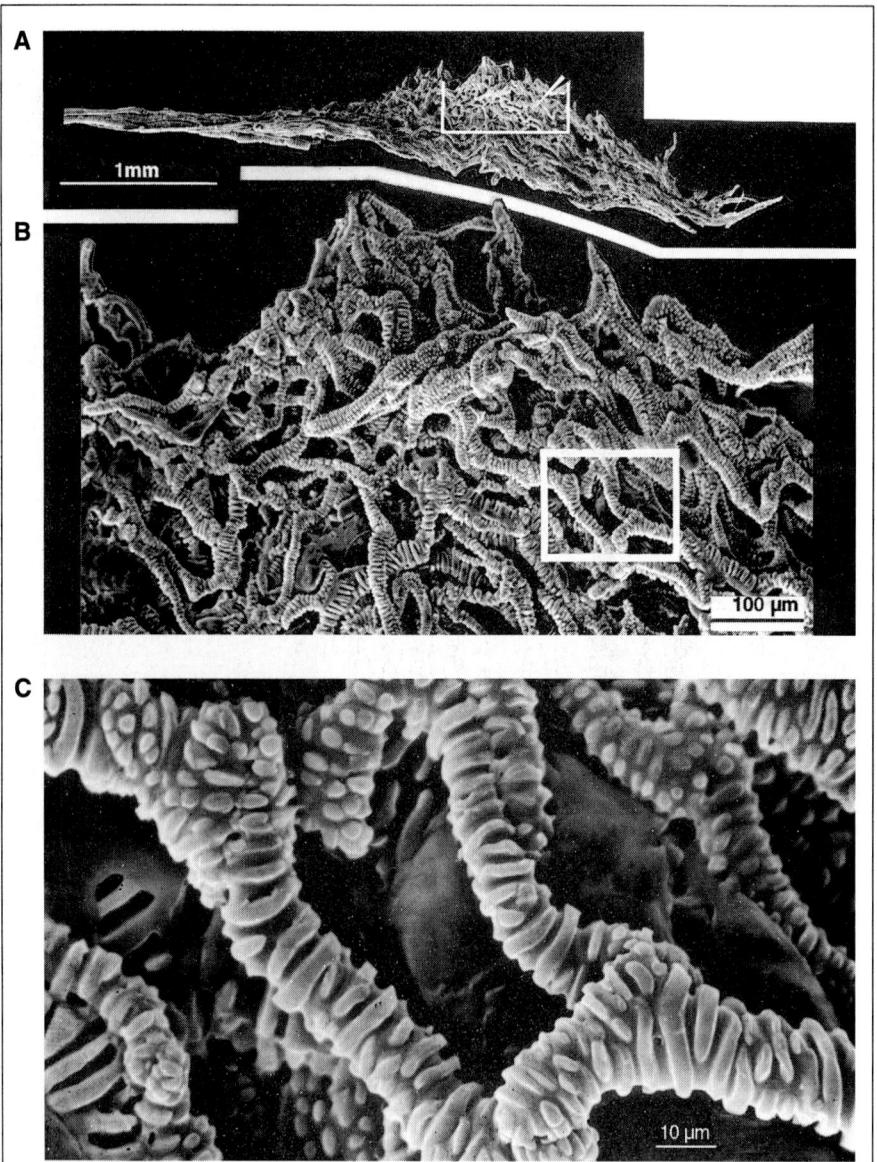

Plate 2.15. Nodal ramification of the foliar vasculature of bamboo *Bambusa tuldoides* (SEM photo). A. A bundle of leaf vein and a nodal bundle join their xylems in two pairs of intricate ramifications such as this. At left are vessels of the nodal bundle. At right are the foliar vessels. The inset and the arrow indicate the location of enlargements B and C respectively. B. Upper central part of A. Annular, reticulate, and pitted elements are distinguished in this inextricable junction. The inset indicates the location of C. C. Enlargement of the inset marked in B: reticulate elements. (See Plate 1.7.)

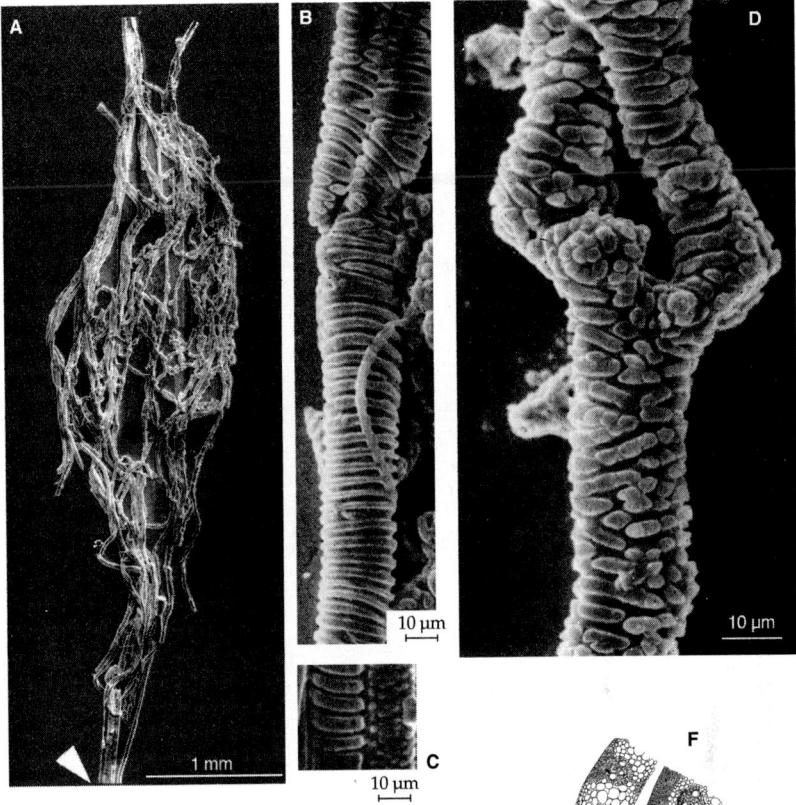

Plate 2.16. Nodal ramification of foliar vasculature in wheat (*Triticum* sp.) (SEM photos). A. Vascular junction between a foliar bundle and a nodal bundle (nodal bundle at bottom). It is composed of helical, reticulate, and pitted elements as in Plate 2.15. B, C, D. Details showing the different types of walls (detail C is located by the arrow in A). E. Nodal segment of wheat showing, in longitudinal section, the position of ramifications belonging to the bundles of the culm (cr) and to the junctions with the foliar vasculature (fr) at the level of this node. The foliar sheath, slit to point b, forms a continuous sleeve below, or *pulvinus* (p) up to the base of its insertion at point a. F. Partial cross-section of the culm and foliar sheath. Location of bundles (px, mx: proto-, metaxylem).

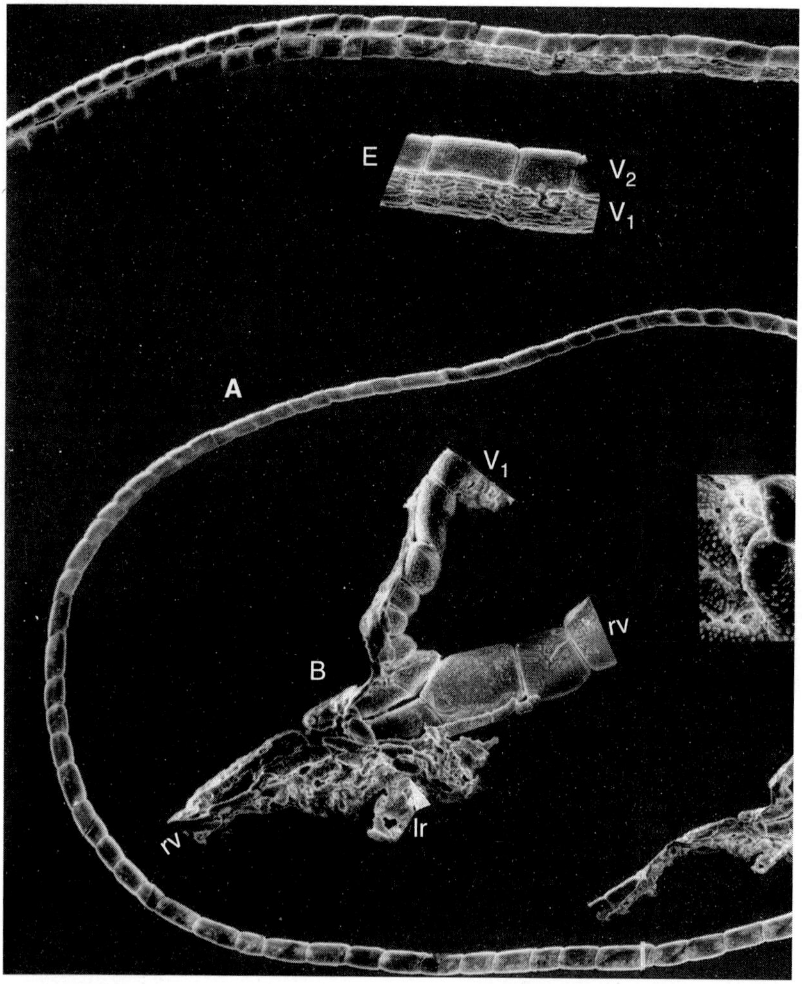
E
V_2
V_1
A
V_1
rv
B
rv
lr

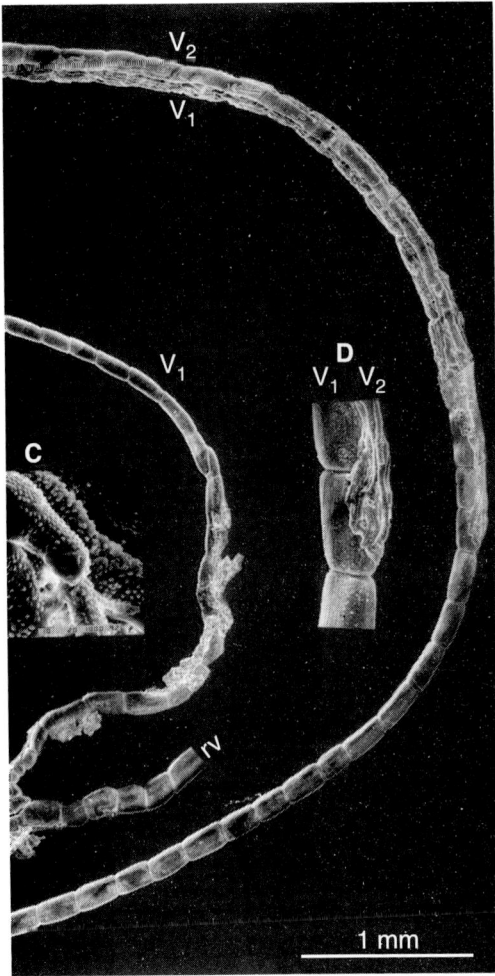

Plate 2.17. The metaxylem in a bundle of rhizome-culm junction of *Phyllostachys viridis* (SEM photo). A. The segment (length 30 mm) is composed of a small portion of the rhizome vessel (rv), of a portion of its lateral branch V_1 (vessel V_1 of Fig. 2.5), and a portion of the second vessel, V_2, in its lower terminal part (V_2 of Fig. 2.5). Partial enlargements: B. Origin of branch. Near the principal branching, there is a small lateral ramification (lr). C. Detail of the lateral ramification (arrow in B). D. Tip of vessel V_2. E. In this portion, V_2 reaches the diameter of V_1. Between D and E, the surface of V_1 is covered by the casts of the adjacent cells.

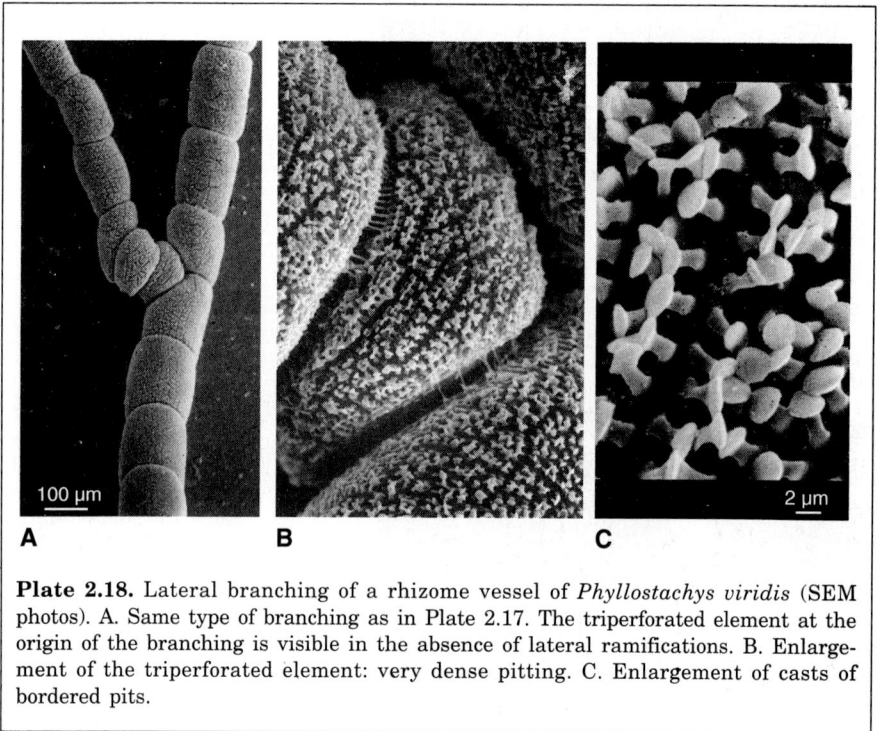

Plate 2.18. Lateral branching of a rhizome vessel of *Phyllostachys viridis* (SEM photos). A. Same type of branching as in Plate 2.17. The triperforated element at the origin of the branching is visible in the absence of lateral ramifications. B. Enlargement of the triperforated element: very dense pitting. C. Enlargement of casts of bordered pits.

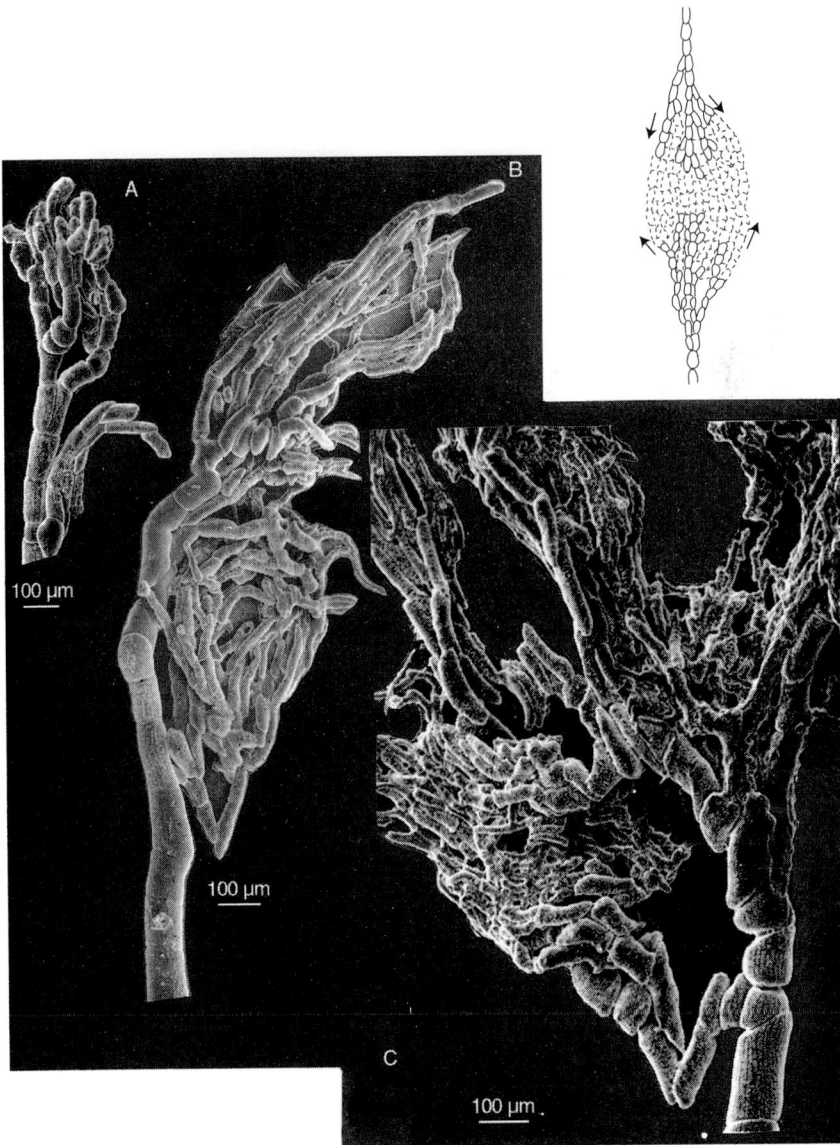

Plate 2.19. Nodal ramifications of metaxylem of *Bambusa tuldoides* (SEM photos). A, B. Ramifications forming in a growing culm undergoing lignification. C. Portion of a ramification in a node of a completely lignified culm. Above right, diagram illustrating the hypothetical extension of two ramifications during the differentiation of their elements.

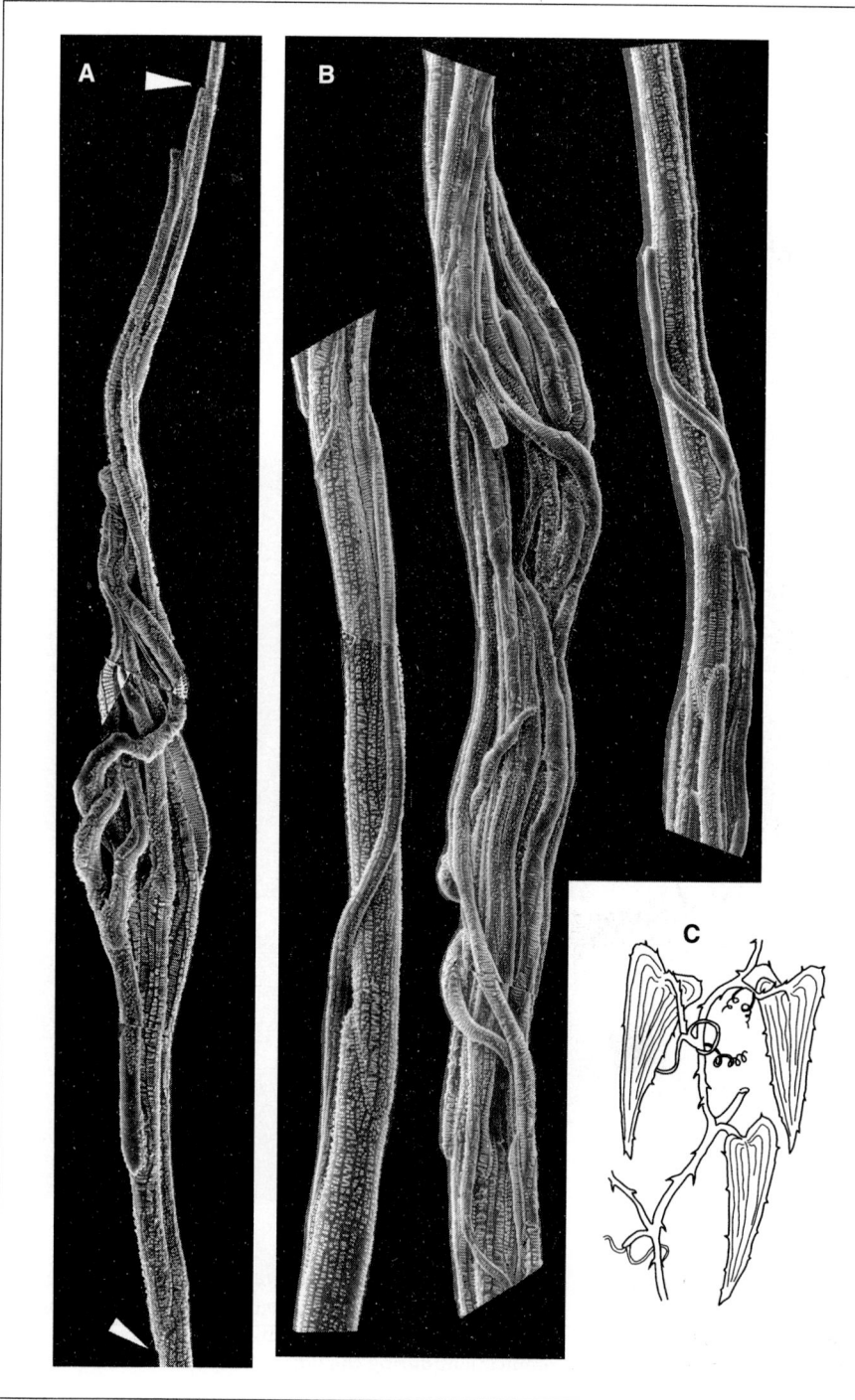

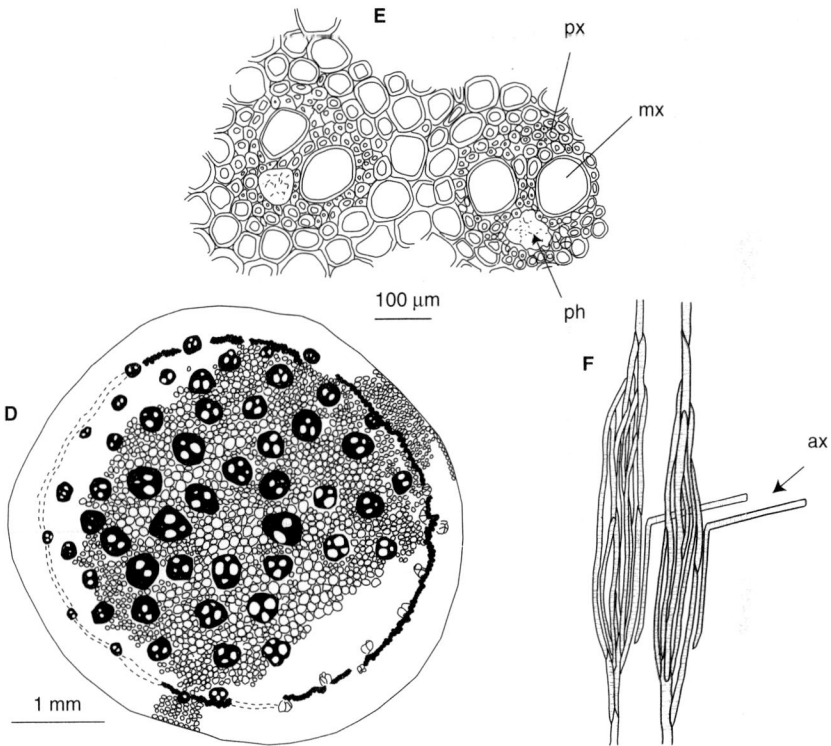

Plate 2.20. Nodal ramifications of metaxylem of *Smilax aspera* (SEM photo). A. Entire ramification. The two arrows indicate the first branching on either side. B. More extended ramification. The cast photograph is divided into three parts, with the lowest part at left and the uppermost at right. The elements have circular and scalariform pits. C. Leafy stem. D. Section of the twig internode: bundles dispersed in a parenchyma. E. Two bundles. F. Diagram of nodal ramifications and of unbranched extremities of vessels of the axillary branch (ax) (px, mx: proto-, metaxylem; ph: phloem).

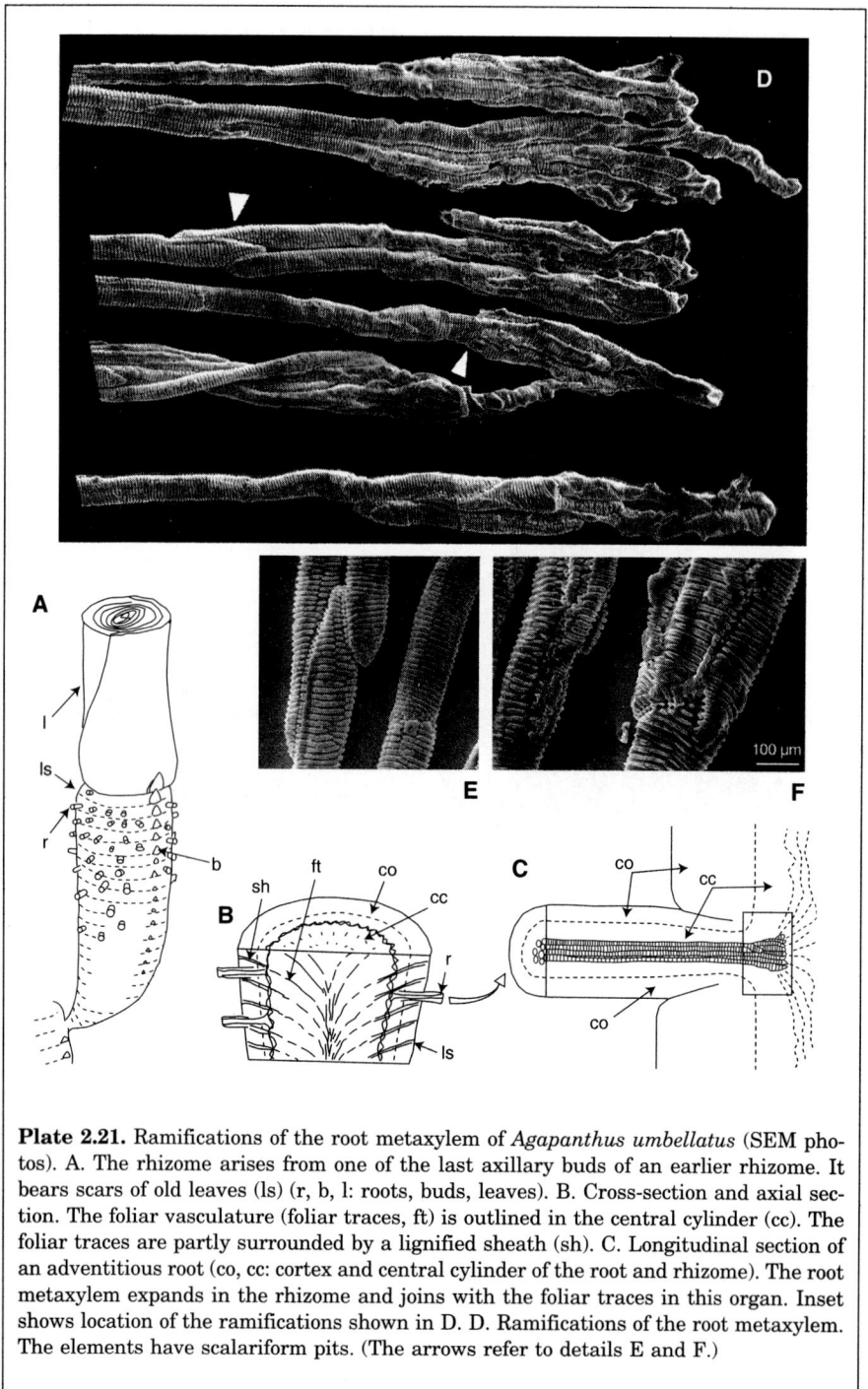

Plate 2.21. Ramifications of the root metaxylem of *Agapanthus umbellatus* (SEM photos). A. The rhizome arises from one of the last axillary buds of an earlier rhizome. It bears scars of old leaves (ls) (r, b, l: roots, buds, leaves). B. Cross-section and axial section. The foliar vasculature (foliar traces, ft) is outlined in the central cylinder (cc). The foliar traces are partly surrounded by a lignified sheath (sh). C. Longitudinal section of an adventitious root (co, cc: cortex and central cylinder of the root and rhizome). The root metaxylem expands in the rhizome and joins with the foliar traces in this organ. Inset shows location of the ramifications shown in D. D. Ramifications of the root metaxylem. The elements have scalariform pits. (The arrows refer to details E and F.)

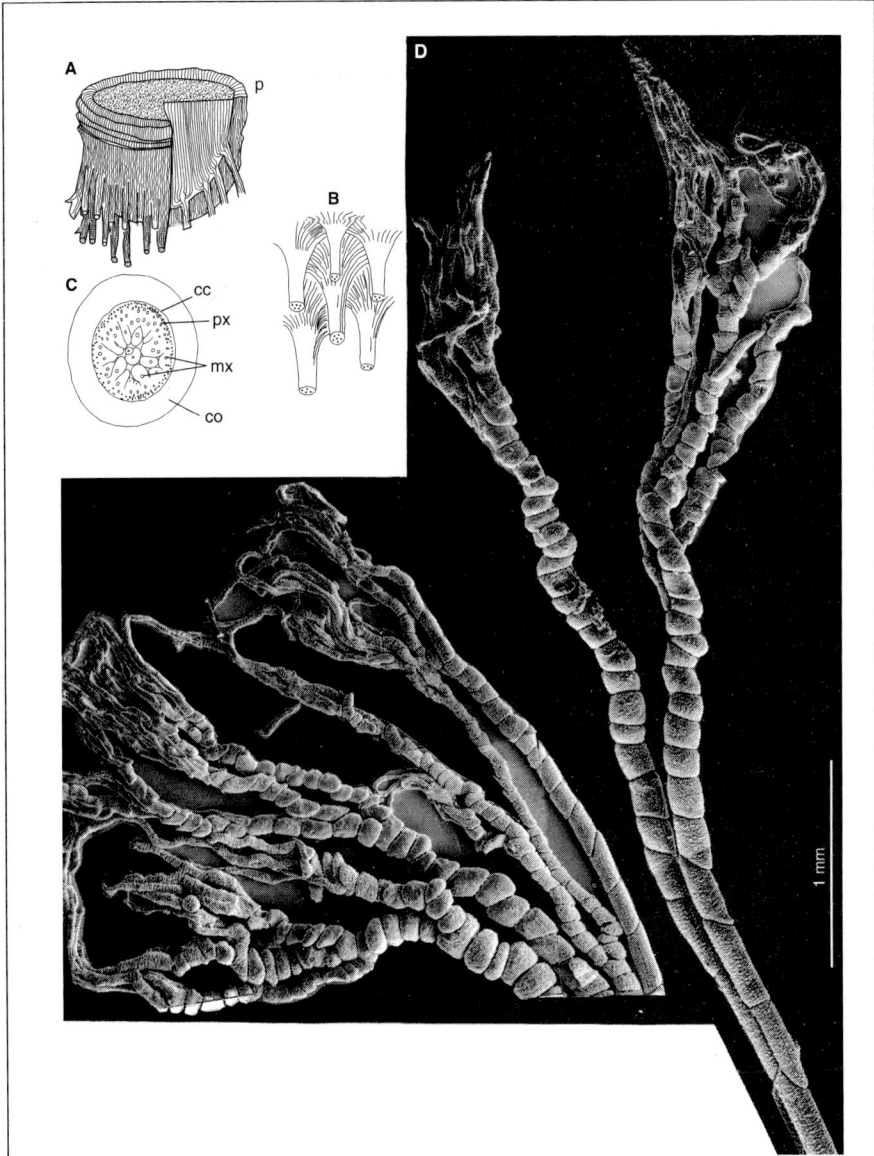

Plate 2.22. Ramifications of the root metaxylem of *Phoenix dactylifera* (SEM photos). A. Base of the partly cut stem of a 10-year-old palm. Vigorous, highly lignified roots rise nearly vertically from the plateau (p). B. The bases of roots form arches in a regular arrangement under the plateau. C. Cross-section of a root. All the cell walls are thick and lignified (co, cc: cortex, central cylinder). D. Ramified extremities of root metaxylem vessels in the plateau. Compare with root metaxylem of maize (Plate 2.14) (px, mx: proto-, metaxylem).

The Secondary Xylem of Dicotyledons
(Intracellular casts)

ADVANTAGE OF VASCULAR CASTING

The secondary xylem (or wood) of Dicotyledons with its four to six constituent cell types is the most complex tissue in terms of differentiation. With its characteristic lignified walls, resistant to changes in weather, it is also a marker tissue in histological terms and a source of morphological data for the identification of temperate and tropical ligneous species, whether living or ancient.

That is not all. In its annual rings the wood records the history of the major events that accompanied its formation, the overall effects of environmental factors, as well as the local constraints and deformations suffered by the cambium during all its periods of activity. Among the wood cells that store this information, the vessel elements in particular constitute cell files of the same age, whose arrangement materializes the wood grain exactly. Moreover, they offer the advantage, from a scientific point of view, of being easy to isolate in each annual ring by means of a casting technique that conserves their continuity and their *in situ* forms. Therefore, the "cambial grain" (of which the wood grain is a copy) is revealed precisely by the casting of vessels of the same generation.

The cambium forms a nearly two-dimensional tissue sheet in each organ, compressed between the already formed wood and the bark. On the basis of observations of the wood grain in the junctions between trunk and branch, and between adjacent organs in general, we suggest the hypothesis that *the cambial sheets of two adjacent organs exert a tangential thrust of growth on each other along their contact edges*, and perpendicularly to this line in each of their points. This hypothesis will be mentioned several times in the following sections.

Casting therefore seems to offer access to new information from the angle of the *histogenesis of the xylem*.

With suitable examples, we will successively tackle the following questions:

— the formation of vessel ends,
— the occasional branching of vessels,
— the effects of cambial constraints and deformations on vascular morphology—*zigzag and circular vessels,*
— the vascular junctions between adjacent axes—lateral and adventitious roots,
— the vascular junctions between a hemiparasite and its host.

FORMATION OF VESSEL ENDS

There are no comparative investigations into the morphology of vessel ends. This problem is partly linked to the study and measurement of lengths of entire vessels, a field that is still largely unexplored (Tyree, 1993).

In casts taken from around 15 species, only vessels with *simple perforations* have *extremities defined* by the following characteristics: (1) the vessel diameter decreases regularly over a length of some tens of elements and (2) there is a uniperforate terminal element, the diameter of which does not exceed about twice that of cambial derivatives. Consequently, vessels with simple perforations appear tapered at both ends (Plates 3.9, 3.10). Among a set of *cast* vascular segments of this type, the vessel ends can be easily spotted and separated from other vessels along which they are closely positioned and to which they are connected by pit pairs (Plate 3.9D).

The tapered shape of these vessels results from the progressive increase in the final diameter of cells differentiating away from the two terminal elements, which implies the precocious determination of the position of the two terminal elements. The gradient of diameter increase of the vessel from its two extremities is certainly linked to a similar gradient of entry of water into the growing cells of the file. The length and position of a developing vessel are thus fixed from the beginning of the differentiation of the file. From the examination of the casts, the vessel extremities do not seem to be linked to the location of nodes along the axes, even in the first annual rings of the wood.

In comparison, vessels with totally or partly scalariform perforations, as for example those of magnolia (*Magnolia grandiflora*) and plane tree (*Platanus occidentalis*), seem to be constituted of unended files of isodiametric elements. These vessels seem to lack ends. In magnolia, the overlapping of elements is significant (Plate 3.11). The cohesion of their casts suggests that some lateral scalariform pits are partly perforated. In the plane tree, scalariform perforations alternate irregularly with simple perforations (Plate 3.12). The arrangement of vessel elements of these two species seems to represent two intermediate stages between the arrangement of tracheids and that of elements with simple perforations of tapered vessels or "true" vessels.

OCCASIONAL BRANCHING OF VESSELS

Vascular branching is rare in the wood of Dicotyledons. To reveal them in a set of casts, the vessel segments are separated by lateral traction until points of resistance indicate the position of triperforated elements at the origin of branching (Plates 3.12, 3.13). The two branches formed are parallel and adjacent over their entire length and are directed either basipetally or acropetally. The elements of each branch close to the triperforated element are often of equal length. It remains to be determined whether the two files are arranged radially or tangentially, i.e., whether they are derived from a single file of cambial initials or from two adjacent files. The second hypothesis seems more probable.

Among the Dicotyledons in which vessels have been examined, the frequency of branching seems higher in the plane tree (Plate 3.12) than in species with simple perforations, such as tomato or walnut (Plate 3.13).

THE EFFECTS OF CONSTRAINTS AND CAMBIAL DEFORMATIONS ON VASCULAR MORPHOLOGY

The vascular structures of *zigzag and circular vessels* that we will compare and describe in this section seem to record, in varied conditions, the deformations undergone by the cambium under the effect of constraints. They suggest reflections and hypotheses on the mechanisms and causes of their formation.

• Zigzag vessels

Zigzag vessels are those in which part of the elements are connected top to bottom by terminal or lateral perforations (André, 2000). The trajectory of a liquid (water, casting medium) through the perforations changes direction while passing from one element to the next in the zigzag. *In situ*, the zigzag elements are joined together into clusters that are difficult to detect in histological sections. Their existence, as revealed in casts, is not mentioned in the recent literature on the wood of Dicotyledons (Carlquist, 1988) and vascular cambium (Larson, 1994), but B. Sundberg has mentioned seeing them in tangential sections of wood (personal communication, 1999). When their casts are stretched out by a slight traction on their ends, they reveal the highly varied arrangements of their elements, which often alternate with long or short linear segments. The continuity of the duct formed by the entire arrangement is proved by the cast itself. These formations have been located especially in nodal zones and in the zone of influence of a hemiparasite, the mistletoe (see later).

—Location in nodal zones (Fig. 3.1)

Zigzag vessels are frequently found in the wood of the first year formed in a shoot, on either side of leaf petiole insertions, and in the wood formed during the second year, on either side of the insertion of axillary shoots. They are found in lignified stems of tomato plants (Plate 3.17) as well as in oak branches one or two years old (Plate 3.15). These vessels are linear in the internodes. Their zigzag portion is located at the place where the insertion of the adjacent organ, petiole or shoot, is the widest. The spring and summer zigzag vessels of oak have the same arrangement.

Such vascular deformations seem to be caused by the growth of adjacent organs; in this case, the tangential thrust exerted by the extension of its own cambial meristem against the cambial files of the axis that bears it might cause simultaneously the overlapping and the shearing of the

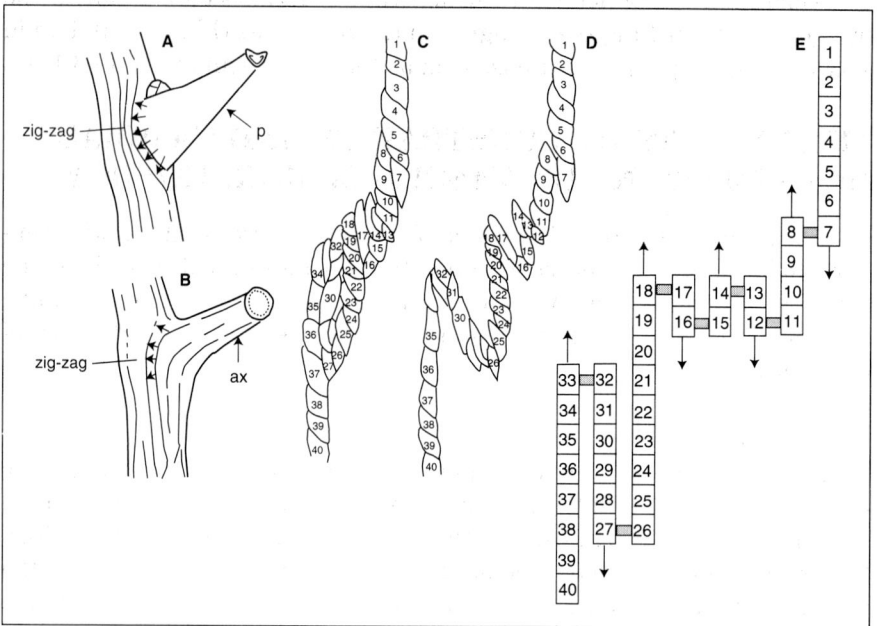

Fig. 3.1. Location and morphology of zigzag vessels. A, B. Location in the nodal zones, on each side of petiole insertions (p) and axillary branch insertions (ax). The arrows indicate the supposed direction of tangential thrusts exerted on the cambium of the axis by the growth of lateral organs. C. Arrangement of the cast in a compact cluster in situ. D. The same is stretched in order to show the vascular continuity (Plate 3.15D). E. Diagram of the same segment broken down into fictional short linear vessels, supposed to be produced from adjacent cambial files that were compressed and "clipped" by lateral thrusts.

latter, as suggested by the analysis of segments of a zigzag vessel of oak (Fig. 3.1). The mechanism thought to be involved might be competition between two adjacent cambial sheets located in the same plane.

—Location in the implantation zone of a hemiparasite (Fig. 3.2)
The extension of mistletoe (*Viscum album*) suckers into the host wood also results in a process of cambial competition, here between two species (Thoday, 1956). In this example, the host is the apple tree (*Malus* sp.). The general direction of the host wood grain and its vessels, sources of water and minerals for the parasite, is deviated towards the suckers through the effect of attractive signals (Fig. 3.2). In this zone of influence, the host vessels develop impressive sequences of zigzag elements (Plate 3.16).

The interaction between the two species has other, even greater, effects on the growth of the host wood and on the establishment of their vascular junctions, which will be examined more thoroughly later (p. 65). It is possible that the cambial competition triggers the diffusion of chemical mediators.

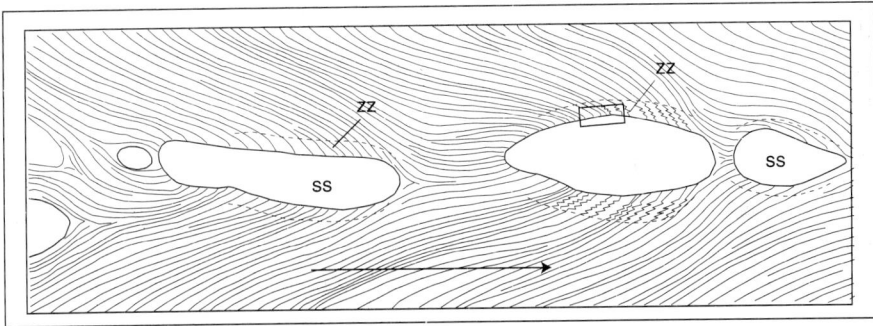

Fig. 3.2. Deviation of the wood grain and vessels of apple tree close to secondary suckers (ss) of mistletoe (*Viscum album*) (illustration based on macrophotograph). Many vessels are thus interrupted on the flanks of the suckers. The vessels form zigzags (zz) in the zones marked with broken lines. The arrow indicates the main direction of water flow in the host wood. The inset corresponds to Fig. 3.8H.

—Location in galls caused by *Agrobacterium tumefaciens* (Fig. 3.3)
In many woody species, *Agrobacterium tumefaciens* induces galls, in which the conditions under which newly formed tissues are structured are totally different from conditions found in normal organs. The secondary xylem that develops in the galls has apparently highly irregular forms, as partly described by Aloni et al. (1995) on galls of castor-oil plant. Vessel casting done on rose gall corroborates and supplements the observations of these authors: in addition to spirals and rings, characteristic zigzag formations are noted (Fig. 3.3 and Plate 3.17).

Remarks: The question arises of the possible physiological advantage, for the leaf, shoot, and parasite, of the formation of zigzag vessels in the

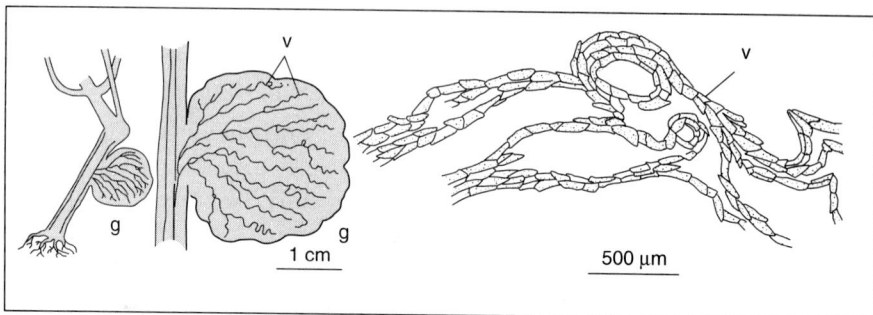

Fig. 3.3. A crown gall (g), developed on a rose stock. Its vasculature (v), connected with that of the stem, is made up of vessels with apparently irregular forms.

stem that bears these organs. If we admit that the secondary phloem follows the same path as the secondary xylem, the question then pertains to both tissues.

The abundance of zigzag vessels at the nodes and the connection of axes could explain entirely or partly first the reduction in hydraulic conductivity that is measured at these points of plant architecture and second the illusion that the tips of xylem vessels are concentrated at these zones (see Zimmermann, 1983).

• Circular vessels

Circular vessels are located in the tangential plane of the wood and frequently form concentric rings.

According to Larson (1994), circular formations were mentioned for the first time by Neef (1914, 1922). Most of our understanding of their morphology and location in the wood is due to the research of Aloni and Wolff (1985), Hejnowicz and Kurczynska (1987, 1991), and Lev-Yadun (1996) following the publication of a study by Sachs and Cohen (1982). Aloni et al. (1995) discovered vessels forming helixes and rings in the galls, as we have mentioned above. These authors, especially Aloni (personal communication), advanced the hypothesis that the formation of circular vessels was initiated by the establishment of self-sustaining circular whorls of auxin movement in the more or less determinate plane of cambium or of its immediate derivatives. In the following lines, we suggest a different hypothesis based on the fact that the circular vessels develop naturally at the intersection of two axes, i.e., along the two orthogonal lines that separate the wood grains of each of them, and are increased in the healing wood that forms on one axis when the other has been cut off (Fig. 3.4).

—Location of circular vessels at the intersection of axes

At the upper part of the junction between a trunk and a branch, the wood grain forms four contiguous sheets, which then merge into two and form

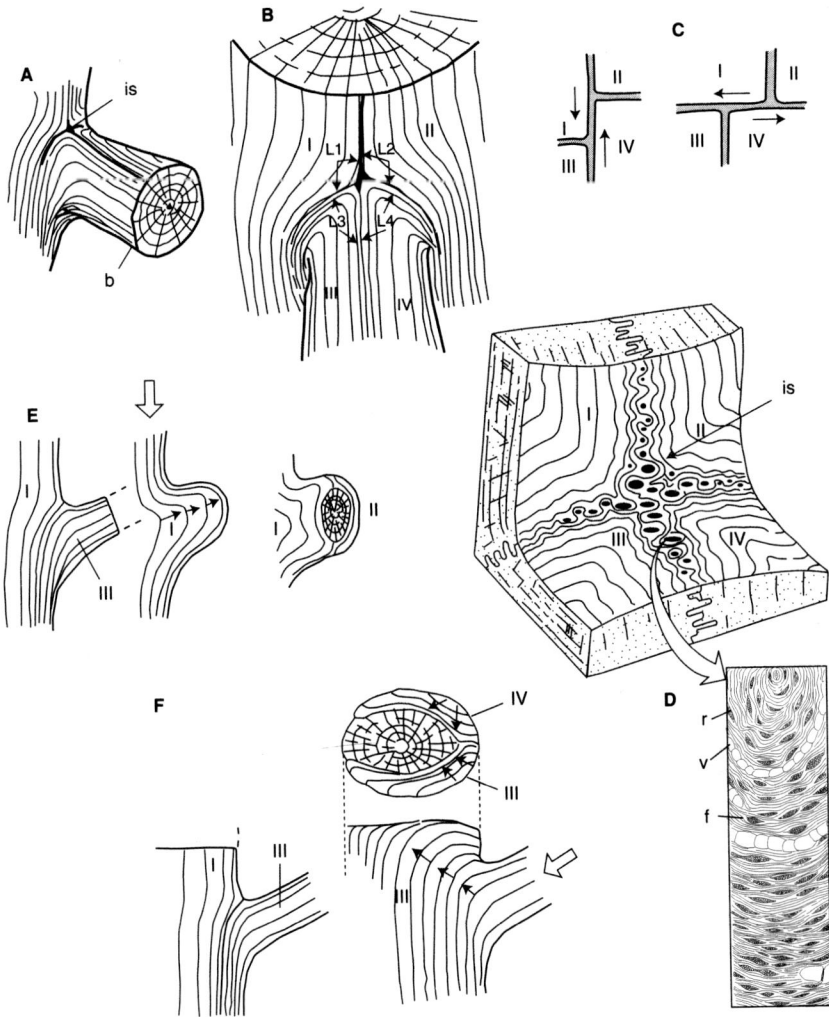

Fig. 3.4. Arrangement of wood grain in the upper part of the junction between a branch (b) and the trunk. A, B. The four edge-to-edge sheets I, II, III, and IV and their edges L_1, L_2, L_3, L_4. The lines are traced in a perfect orthogonality and form an intersection (is). C. In reality, the lines of junction are frequently shifted and their trails are irregular. D. Formation of circular zones near the intersection. All the xylem cells, vessels (v), fibres (f), and rays (r) are arranged in concentric circles. E. Extension of sheets I and II after the pruning of the branch. The large arrow indicates the direction of metabolite flow. F. Extension of layers III and IV after the decapitation of the trunk.

the cylinders of the two axes (Fig. 3.4). This morphology is the same as that at the junction of any two branches. The wood sheets are produced by cambial sheets of the same form. These four cambial sheets are believed to exert one over the other tangential thrusts along their adjacent borders L_1, L_2, L_3, L_4. These contiguous borders form two more or less orthogonal lines the positions of which, year after year, are determined by the fluctuating equilibrium between the thrusts, which is linked to the metabolic and hormonal flows supplying each sheet. The intersection often vanishes because of the frequent shift in the lines of junction (Fig. 3.4A, C).

Near these lines, the wood grain nearly always forms undulations with loops including circular formations with a diameter of only a few millimetres in the first rings up to a few centimetres in the old wood. In the circular formations, the wood cells, fibres, vessel elements, and parenchyma are arranged in concentric lines (Fig. 3.4D). It should be noted that the wood of conifers has comparable formations: both fibres and vessel elements are missing but tracheids and radial parenchyma are present.

In circular formations, the vessel elements are assembled in concentric annular loops. The question of their functional utility has naturally been raised. It has been hypothesized that they play a role in limiting water circulation from the trunk towards the branch in case of stress (Lev-Yadun and Aloni, 1990). However, some observations do not conform to that idea. Since most of the ascending sap flows through vascular pathways located on the lower side of branches, when the circular formations develop on the upper side, we estimate that this limiting role must be negligible. Moreover, there are good reasons to suppose that the water moves tangentially across circular vessels via pits. Finally, it should be added that circular vessels are also formed at the junction of the main roots.

—Location of circular vessels in healing wood
The pruning of a branch breaks the equilibrium of the metabolic flows from the leaves and hormonal flows from the apices and consequently breaks the equilibrium of the thrust of the four cambial sheets. This occurs immediately or over time, depending on the season in which the tree is pruned. The cambial sheets of the pruned branch die and are progressively covered by those of the trunk in a healing process. In the same way, the decapitation of the trunk leads to a similar production of healing wood by the branches, if they are vigorous enough (Fig. 3.4E, F). For example, let us examine the healing wood formed by the main branch of a eucalyptus (*E. globulus*) around the base of a pruned branch (Fig. 3.5A, B).

On the flanks of the block of wood cut from the main branch (Fig. 3.5B, C, D), the wood grain marks the progression of two healing "rims". The rims advance by anticlinal divisions of the extreme edges of their

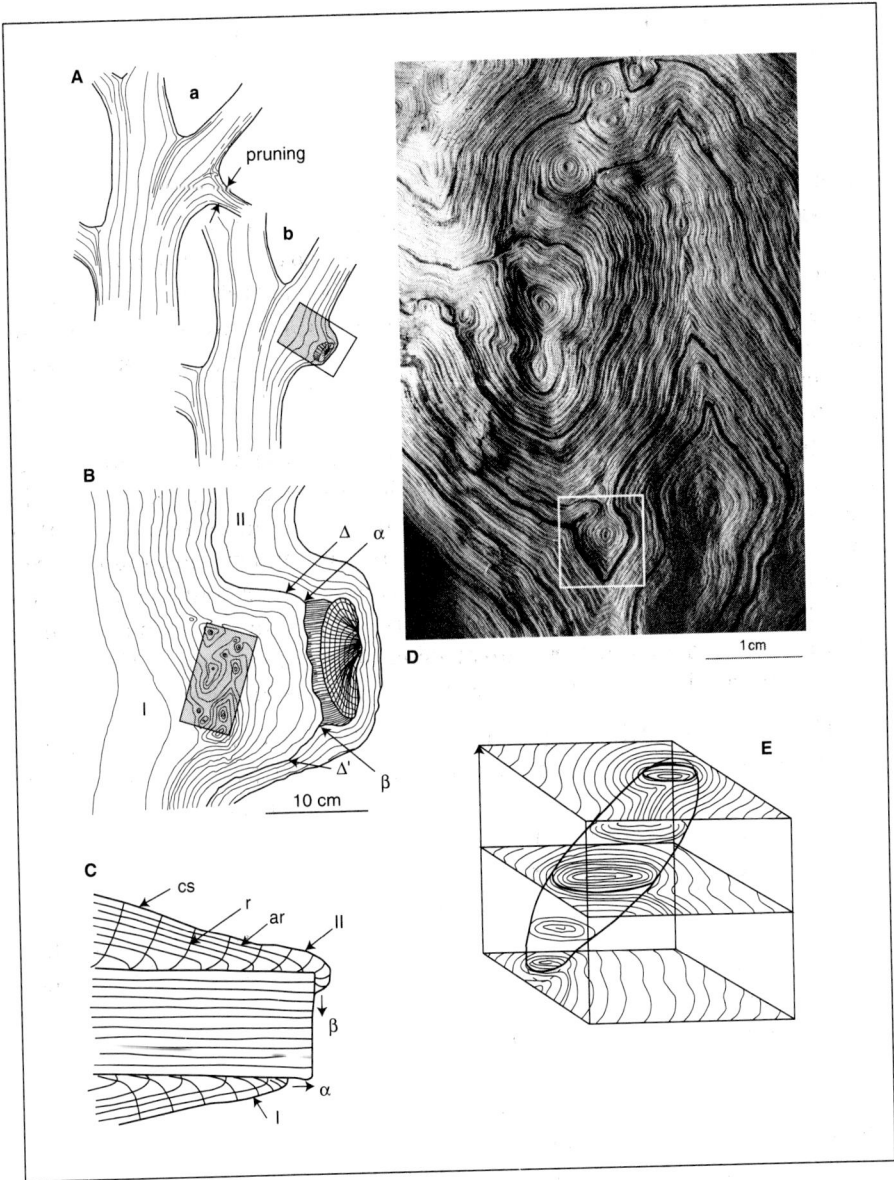

Fig. 3.5. Circular vessels formed in the healed wood. A. Eucalyptus (*E. globulus*) trunk at the time of pruning (arrows) (a) and around 10 years later (b). In the rectangle is the block of wood examined. B. The block of wood. Extension of the healing rims of the two sheets, I and II, and their lines of junction Δ and Δ'. Formation of circular vessels especially on flanks of sheet I (inset). C. Cross-section of the two healing rims on either side of the dead wood. The curvature of the rays (r) is always orthogonal to the limits of the annual rings (ar) and to the cambial surface (cs). D. Detail of inset in B. Concentration of circular vessels on the flank of I. The circular vessels of Plates 3.18 and 3.19 come from the part shown in the inset. E. Diagram of a "vortex" in the thickness of the wood.

cambium: they are extensions of the preceding layers I and II, which cover the dead cambium of layers III and IV of the pruned branch. In the wood of the two rims, examined in cross-section (Fig. 3.5C), the rays and the limits of annual rings remain in perfect orthogonality up to the extreme edge. The earlier limits L_1 and L_2 of the sheets I and II come into contact along the lines Δ and Δ' up to α and β respectively (Fig. 3.5B).

On the lateral sides of the healing rims, the wood grain presents numerous circular formations (Fig. 3.5D). In this part of the wood, a series of small plates of wood of 2 mm thickness have been carefully sawn parallel to the surface to a depth of 20 mm. They reveal that the circular formations evolve over time, deform, move, appear in some places, and disappear in others. Figure 3.5E suggests the spatial form of one of these immobile "vortexes".

The vessel casting was done on each plate after the circular vessels were opened by a small scalpel incision. Surface casts of the part shown in the inset in Fig. 3.5D are presented as examples in Plates 3.18 and 3.19. A series of nine concentric circular vessels is seen under scanning electron microscope (SEM) on one side (Plate 3.18A) and under light microscope on the other side (Plate 3.19). The series was inserted in the open concentric vascular loops visible under transmission microscope (Plate 3.18C).

The major observations can be summarized as follows:

— The smallest of the open loops (Plate 3.18C) is concentric with the largest of the closed loops (Plate 3.18A). Apart from the open/closed character, the continuity between the two sets is perfect.
— Vessel elements, fibres, and parenchyma cells cast at the same time (visible on Plate 3.18B) are identical in the closed and open loops.
— Finally, and this point is very important, vessel ends are present in the open loops and in the closed loops (see Plate 3.19A, B) with comparable frequency: in cast vessel segments, taken partly from open loops and partly from closed loops, and totalling respectively 134 cm and 36 cm, i.e., around 6700 and 1800 vessel elements, the vessel tips amounted to 8 and 3 respectively.

— Hypothesis on the formation of circular vessels (Fig. 3.6)
The term "cambial front" is used here to describe the extreme edge of the cambium of a healing rim that advances by anticlinal divisions on the surface of the dead wood. It is presumed that (1) this front is never straight but has an undulating contour with convex and concave parts and (2) the anticlinal divisions are more frequent on the convex parts (André, 2000).

In these conditions, the curvature of the front loops increases during the progression of the front, until the loops touch each other and isolate islands within which the radial and fusiform initials of the cambium form

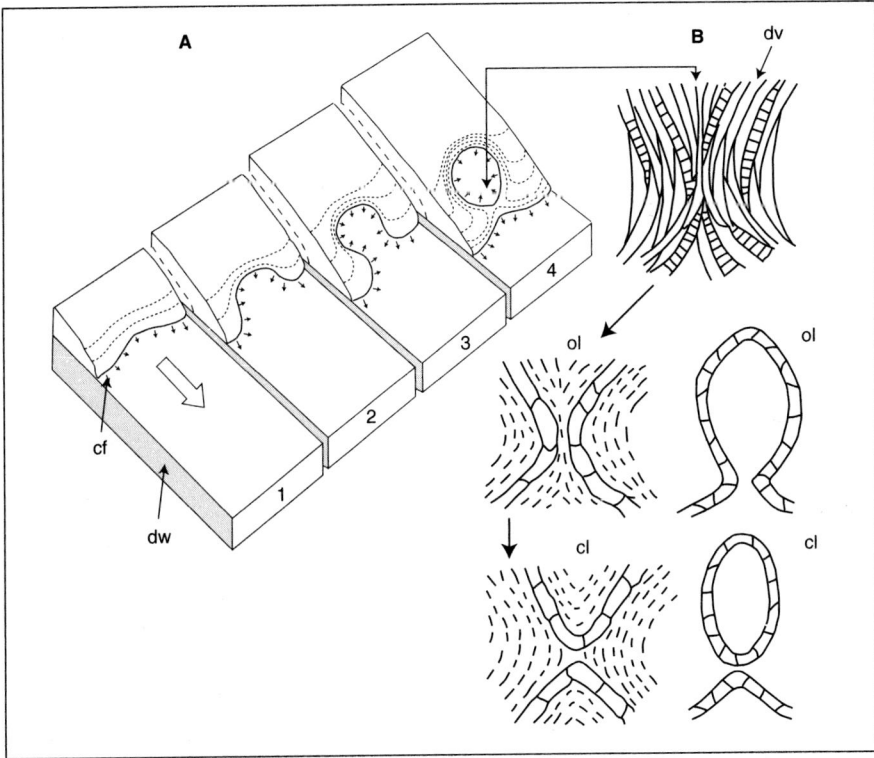

Fig. 3.6. Diagram illustrating the hypothesis of the formation of circular vessels. A. progression, by anticlinal cell divisions, of the cambial front (cf) on the dead wood (dw) with an accentuation of its curvature (stages 1, 2, 3) until the closure of the loops (stage 4). B. A local inversion in the contacts between the differentiating vessel elements (dv) leads from the open loop (ol) to the closed loop (cl).

concentric circles as they become enclosed. From these are derived circular formations of xylem (and phloem?). Continuing its progression, the cambial front loses its excessive curvature and takes on a more linear profile and a subsequent cycle of increase in curvature (Fig. 3.6). On the healing rims that we just examined (Fig. 3.5B, D), the wood grain effectively regains downstream of the circular zones the trace of a slight curvature that it had upstream; but this observation is not the proof on which the hypothesis illustrated in Fig. 3.6 can be founded.

VASCULAR JUNCTIONS BETWEEN ADJACENT AXES

In the secondary xylem of Dicotyledons, vessels are formed either entirely in the wood of one axis or straddle two adjacent axes (Fig. 3.7). In fact, the

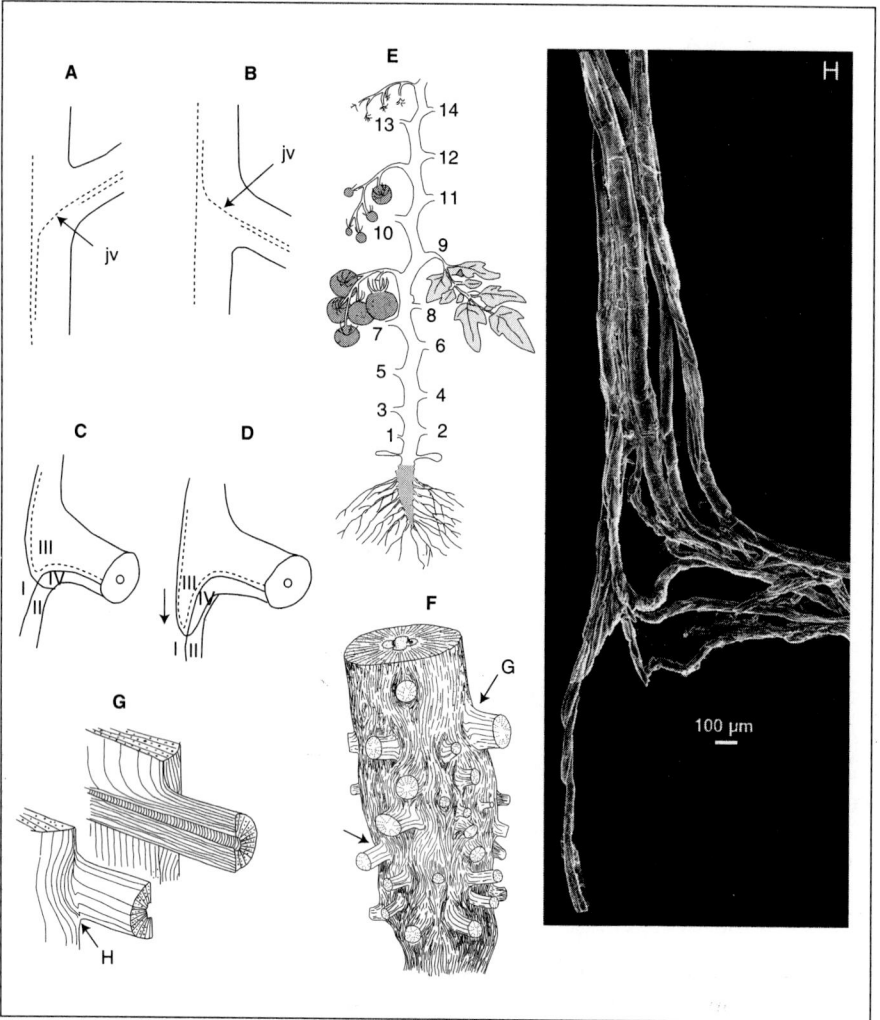

Fig. 3.7. Secondary junction vessels (jv) at the junction between adjacent axes. A. Junction vessel with an axillary branch. B. The same with a root. C. The secondary xylem of roots forms sheets I, II, III, and IV at the junctions as in the aerial system. D. Sheets III and IV often stretch longitudinally. This frequently occurs in junctions between lateral roots and taproot of tomato, both of which have secondary xylem. E. Root system of tomato. F. Detail of E, main part of the lignified taproot with lateral roots removed. The wood grain is apparent in the taproot. G. Detail of F, lateral root sectioned. The direction of xylem vessels is indicated. H. Cast of some junction vessels in the taproot and lateral root of tomato arranged as *in situ* (SEM photo).

position of vessel ends does not seem to be linked to the particular position of branches of the axes, according to an examination of vessel casts. This is true for the aerial and root axes. By means of intervascular pits, the vessels that extend between two axes are in contact with the vessels belonging to each axis, between which they ensure the continuity of water transport (Cruiziat, 1984). These are here called "junction vessels" (Fig. 3.7A, B).

The root vascularization is less thoroughly studied than that of the aerial parts. Therefore, to illustrate the morphology of junction vessels, we point out two examples of the secondary vascularization of lateral roots in tomato plant and adventitious cauline roots in *Ipomoea learu*.

The wood grain forms four sheets on the bottom side of a lignified root junction (Fig. 3.7C, D), comparable to those that form on the upper side of a branch junction (Fig. 3.4A, B). This structure of the wood is visible on the surface of the taproot and lignified lateral roots of the adult tomato plant (Fig. 3.7E, F, G). The casts of some junction vessels of a lateral root of tomato are arranged on the same plane for the photograph in Fig. 3.7H. Their curvature depends on their position in the root as well as the more or less stretched-out form of sheets III and IV (Fig. 3.7D).

The adventitious roots of the stem of *Ipomoea* have junction vessels of similar form. Those that follow the outer edge of the sheets III and IV have a characteristic U shape (Plate 3.20).

Remark: The junction vessels of lateral and adventitious roots have an acropetal direction in the main axis, taproot, and stem, so that the transported water flow is directed towards the apex, unlike the junction vessels of branches, which have a basipetal direction in the stem (Fig. 3.7A, B).

VASCULAR JUNCTIONS BETWEEN A HEMIPARASITE AND ITS HOST

In this section we will discuss the vascular continuity established between the tissues of two different species, a hemiparasite, the white mistletoe (*Viscum album*), and its dicot or conifer host. From this particular junction, the cast gives not only a novel image, but also a more comprehensive view than that obtained by conventional histological techniques.

The white mistletoe, common in Europe, has three subspecies, one parasitizing deciduous trees, two others parasitizing conifers. It meets its water and mineral needs by diverting part of the ascending sap of its host, but it seems to satisfy its needs for carbon metabolites through its own photosynthesis. Its partial dependence classes it among the hemiparasites (Sallé et al., 1993).

Ascending sap is transferred between the secondary vascular systems of the host (vessels or tracheids) and parasite (vessels) by a large number of chains of conducting elements (pseudo-vessels) differentiated in the tissues of the parasite according to processes that have not yet been completely elucidated. Since the pseudo-vessels are made up of perforated elements, the vascular continuum can be completely cast in the case of a mistletoe-apple pair and in the case of a mistletoe-fir pair.

• Vascular junction between mistletoe and apple tree (*Malus* sp.) (Fig. 3.8)

Let us examine a segment of the parasitized branch (Fig. 3.8A). When the bark is superficially peeled, it is apparent that the base of the main stem of the mistletoe extends upstream and downstream inside the host bark by extensions in the form of thick strands, the cortical strands (Fig. 3.8C). The secondary xylem formed in the cortical strands is continuous with that of the mistletoe stem and produced by the extension of the cambium of the stem.

When the host wood is barked and the cortical strands are excised to the base of the stem, small, deep cavities are found in the host wood under the strands (Fig. 3.8C, D). These wedge-shaped cavities radially traverse some annual rings of the host wood and are filled with radial cell files produced by the cortical strands: they are secondary suckers of the parasite.

The stem wood of the mistletoe also extends into the host wood but deeper in the form of a large wedge and constitutes the main sucker or primary haustorium.

It is remarkable that the limits of annual rings of the host and parasite wood coincide exactly with one another. This is evidence that the wood layers of the two species have been deposited simultaneously since the germination of the mistletoe seed. The cambium of the parasite remains positioned constantly, edge to edge, at the level of the host cambium; it inhibits the extension of the host cambium and extends a little further each year at its expense. The passive implantation of the primary and secondary suckers is the result of cambial competition without destruction of the host wood.

The effects induced by the parasite on the host wood tend to prove that the cambial activity of the host depends on that of the parasite, as indicated by the following:

— The branch produces rings that are wider on the side where the parasite grows than on the other side (Fig. 3.8E).
— The host wood grain, as a copy of the host cambium, is deviated in such a way that a large number of host vessels run not only into the flanks of the secondary suckers, where they are interrupted, but also in continuation of the vessels of the primary haustorium. Moreover,

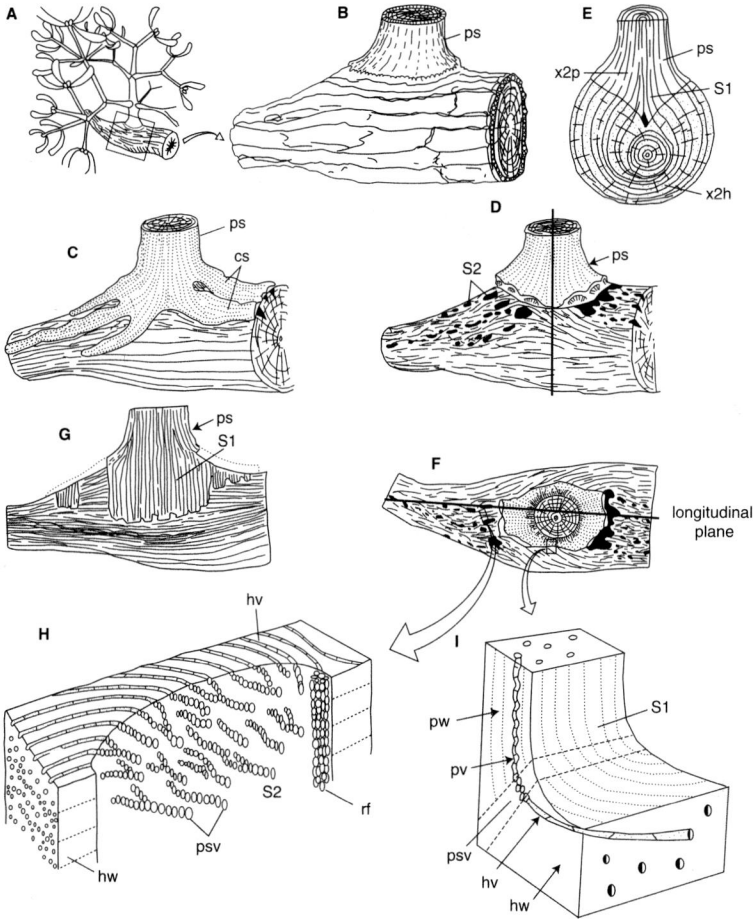

Fig. 3.8. A. Implantation of mistletoe on a branch of apple tree. B. Base of parasite stem (ps) and host branch, bark intact. The sap flows from right to left. C. Same organs with bark peeled: the extensions of the parasite or cortical strands (cs) in the host bark are revealed. D. Same organs after severing of cortical strands. The strands contain a cambial arc and primary and secondary conducting tissues continuous with those of the parasite stem (ps). At the points of contact of the strands with the host cambium, "secondary" suckers develop (S2). The section plane (vertical line) refers to E. E. Cross-section showing the haustorium or "primary" sucker (S1) forming the base of the parasite stem. From the time of seed germination, the host cambial activity is conditioned by that of the parasite (see text). The wood of the two species grows with the same annual rhythm. F. View of D from above; the section plane (horizontal line) refers to G and the two insets to H and I. G. Longitudinal section showing in this plane the orthogonality between the host wood grain and the direction of cell files of the parasite. H. Radial section of a secondary sucker showing radial files (rf) of parenchymatous cells. The host vessels (hv) that end on the flanks of the sucker are connected to pseudo-vessels (psv) differentiated at the same time and densely pitted (see Plate 3.21 and Fig. 3.2). I. Block of tissue at the junction between host wood (hw) and parasite stem wood (pw). The host vessels (hv) and parasite vessels (pv) are connected in S1 by short pseudo-vessels (psv) that ensure vascular continuity (see Plate 3.22).

we have already mentioned the zigzag arrangement of the host vessel elements close to the secondary suckers.

— While the host cambial derivatives become differentiated into vessel elements, parasite cells that come into contact with them differentiate by induction into perforated and pitted conducting elements, with lignified secondary walls (Smith and Gledhill, 1983). Through step-by-step, progressive differentiation, they form continuous chains of some tens of elements, called *pseudo-vessels* by Launay (1950). The pseudo-vessel elements are ovoid and have their major axes radially oriented and then perpendicular to the vessel elements of the host. Their general form is that of parenchyma cells in radial lines that fill the *secondary* suckers and among which they differentiate. The flanks of secondary suckers are carpeted with pseudo-vessels, illustrated in the cast (Fig. 3.8H and Plate 3.21). The pseudo-vessels are oriented towards the centre of suckers, where they terminate. The water is brought by the pseudo-vessels within the radial files of living cells of the parenchyma, which transport it (in an apoplastic or symplastic way) up to the vessels of the cortical strands. Numerous groups of pseudo-vessels have often common elements (see caption of Plate 3.21). In the *primary* sucker, the pseudo-vessels link the secondary vessels of the host and those of the parasite (Fig. 3.8I and Plate 3.22). The element files thus constituted are entirely perforated and the water circulates in them as in single vessels, up to the aerial parts of the parasite.

— When the foliage of the mistletoe is developed, it extracts so much water that the host branch dries beyond the implantation of the parasite.

• Vascular junction between mistletoe and fir (*Abies alba*)

In its implantation and its effects on its host, the subspecies of mistletoe associated with fir has great similarities with that associated with apple. The main difference observed is that the wood grain of the fir does not seem to be deviated towards the suckers, unlike in the apple tree. We suggest the following explanation: the diversion of the water flux is "useless" when it circulates not in a few vessels, but in a diffuse manner in a large mass of connected tracheids. The transport continuity of water between the tracheids of the host and the secondary vessels of the parasite via the pseudo-vessels can be established from the casts (Plate 3.23). Also, the casts reveal a morphological difference between the pseudo-vessels and those of the preceding species, but do not allow a clear analysis of their mode of junction with the tracheids.

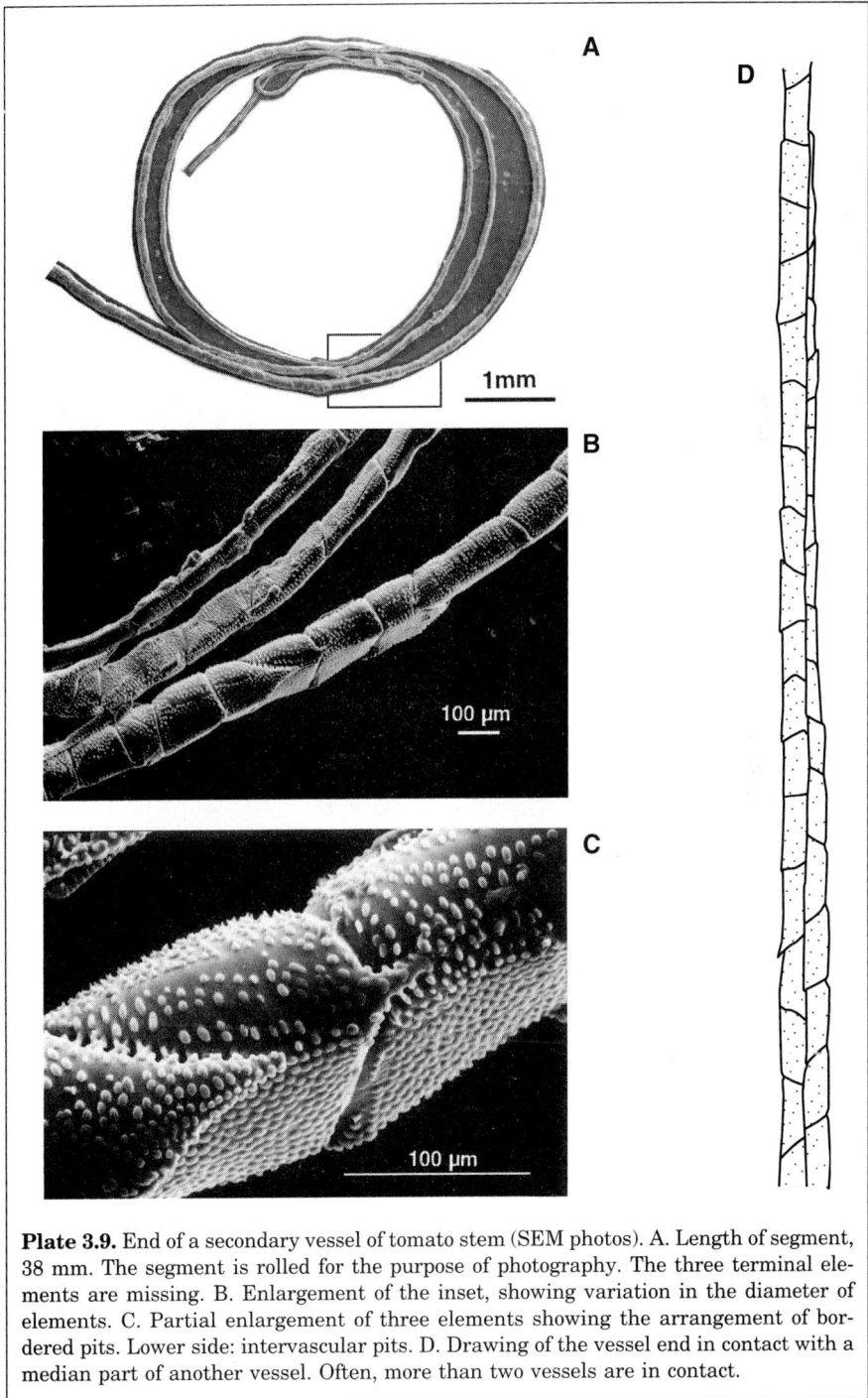

Plate 3.9. End of a secondary vessel of tomato stem (SEM photos). A. Length of segment, 38 mm. The segment is rolled for the purpose of photography. The three terminal elements are missing. B. Enlargement of the inset, showing variation in the diameter of elements. C. Partial enlargement of three elements showing the arrangement of bordered pits. Lower side: intervascular pits. D. Drawing of the vessel end in contact with a median part of another vessel. Often, more than two vessels are in contact.

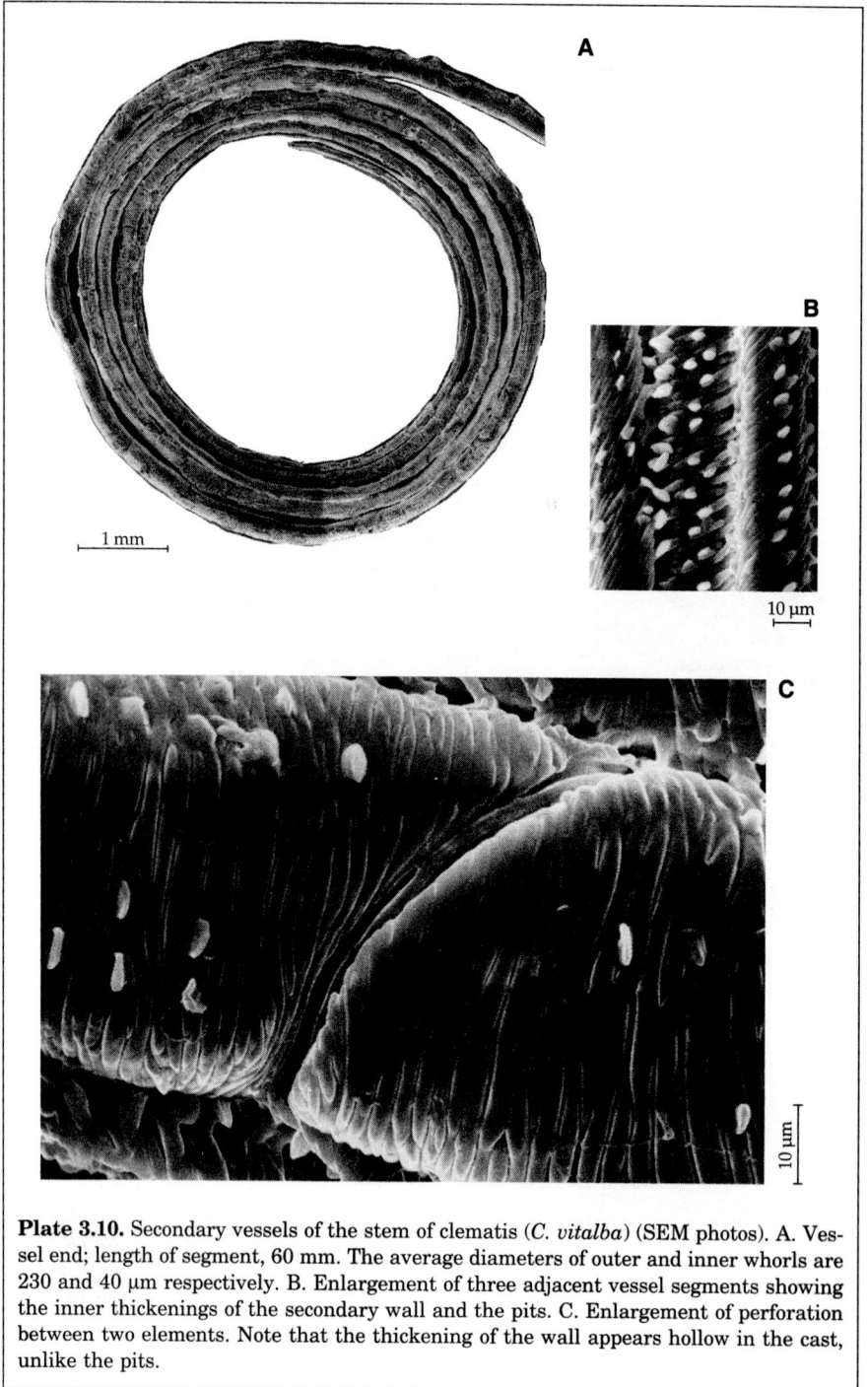

Plate 3.10. Secondary vessels of the stem of clematis (*C. vitalba*) (SEM photos). A. Vessel end; length of segment, 60 mm. The average diameters of outer and inner whorls are 230 and 40 µm respectively. B. Enlargement of three adjacent vessel segments showing the inner thickenings of the secondary wall and the pits. C. Enlargement of perforation between two elements. Note that the thickening of the wall appears hollow in the cast, unlike the pits.

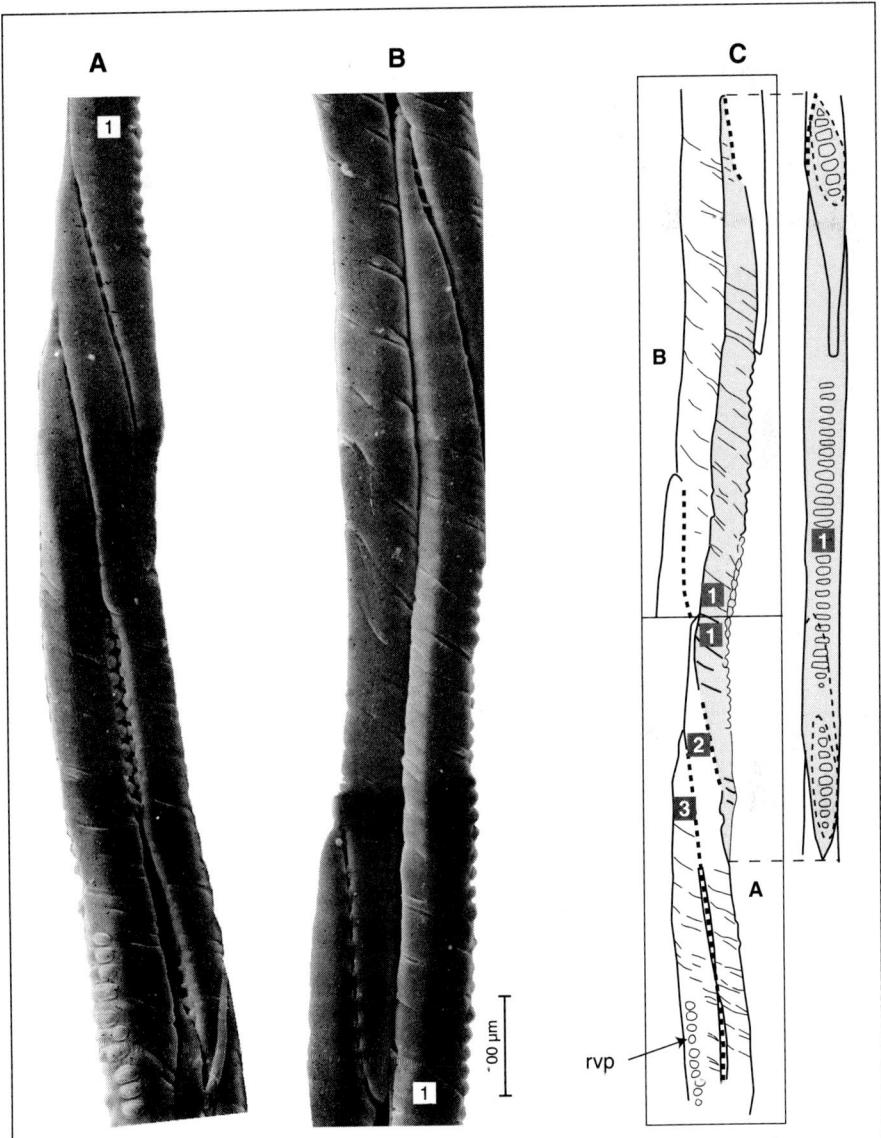

Plate 3.11. Vessel elements of magnolia stem (*M. grandiflora*) (SEM photos and drawings). A, B. The photograph comprising the entire element 1 is in two parts (reconstructed in drawing C). The elements (average length and diameter 600 and 50 μm respectively) are vermiform, with flat tips (drawing C). The files of elements, cast via oblique scalariform perforations, seem uninterrupted, lacking terminal elements. Ray-vessel pits (rvp) are visible on element 3. The helical thickenings on the inner surface of the secondary wall have the same left-hand direction among the numerous elements of this stem sample. C. Drawing of connection of the two parts A and B. Front and side view of element 1 in its entirety.

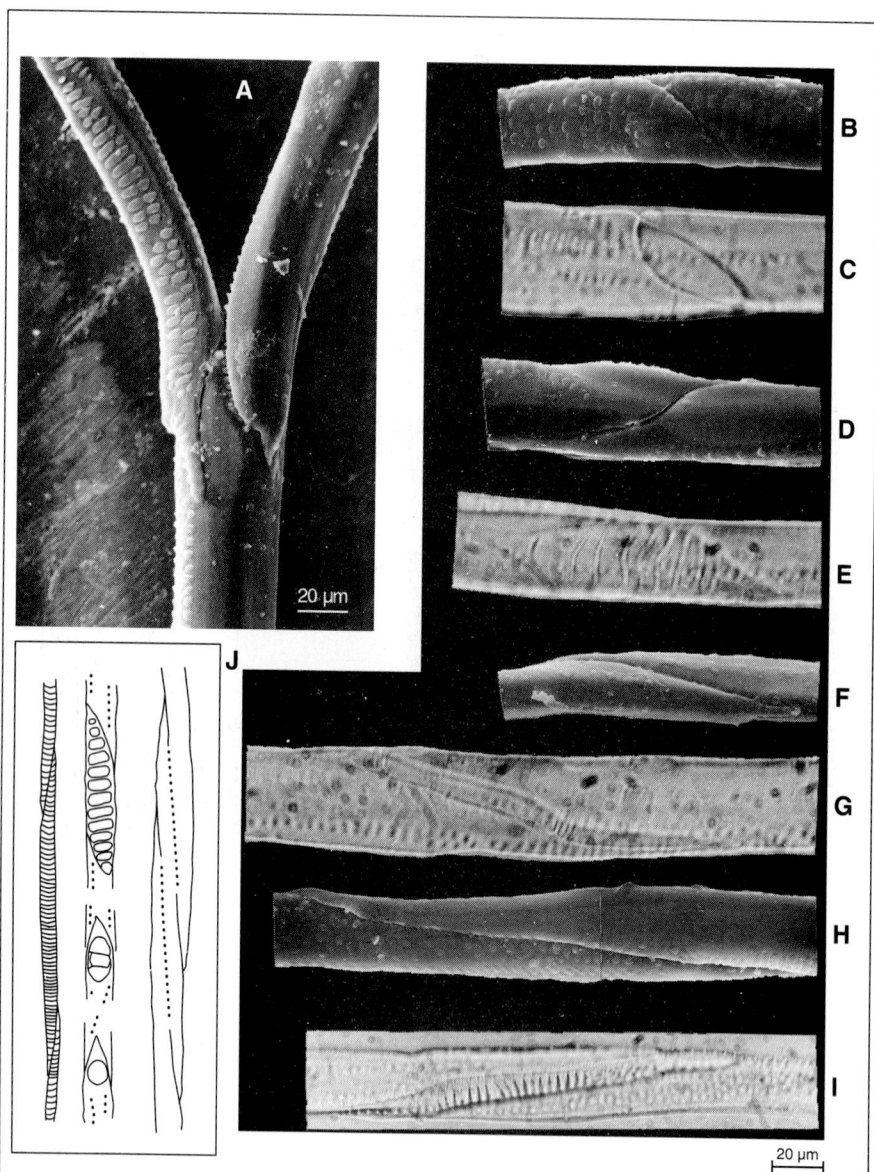

Plate 3.12. Vascular segments of plane tree (*Platanus occidentalis*) (A, B, D, F and H are SEM photos. C, E, G and I are transmission light photos). A. The casts reveal the existence of many branched vessels. The two parallel branches have been spread out for the photograph. In a single vessel, the perforations are either simple (B, C) or scalariform, with a few bars (D, E) or many (F to I). The pits are small. J. From left to right: drawing of an element of the primary xylem; front view of simple and scalariform perforations; side view of significant overlapping of three elements with scalariform perforations in secondary xylem.

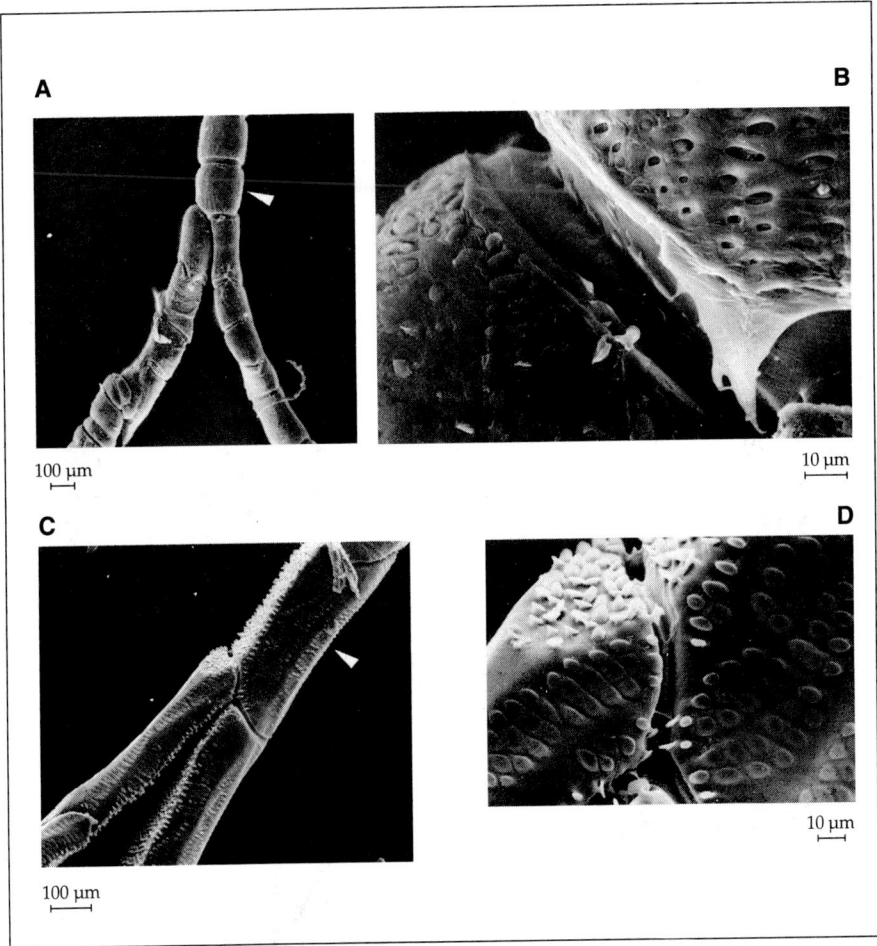

Plate 3.13. Branched vessels of tomato (A, B) and walnut (*Juglans regia*) (C, D) (SEM photos). A, C. The triperforate elements are indicated by an arrow. The two branches, here spread out, are in contact *in situ* along their entire length. B. Enlargement of the "supernumerary" perforation of A; the secondary wall remains, partly detached. D. Enlargement of the "supernumerary" perforation of C. Ray-vessel pits seen in front view. Above, intervascular pits seen in three-quarter view.

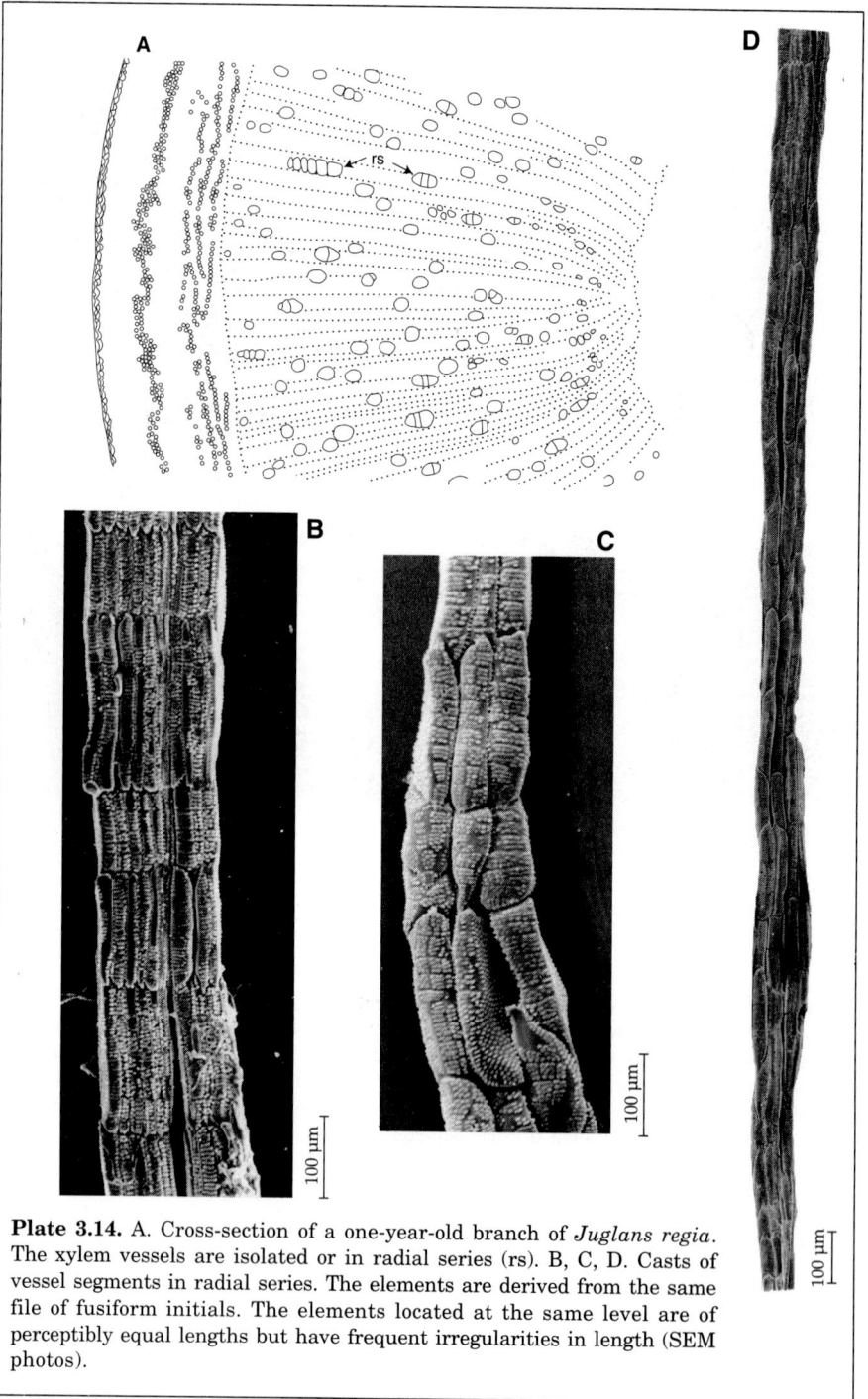

Plate 3.14. A. Cross-section of a one-year-old branch of *Juglans regia*. The xylem vessels are isolated or in radial series (rs). B, C, D. Casts of vessel segments in radial series. The elements are derived from the same file of fusiform initials. The elements located at the same level are of perceptibly equal lengths but have frequent irregularities in length (SEM photos).

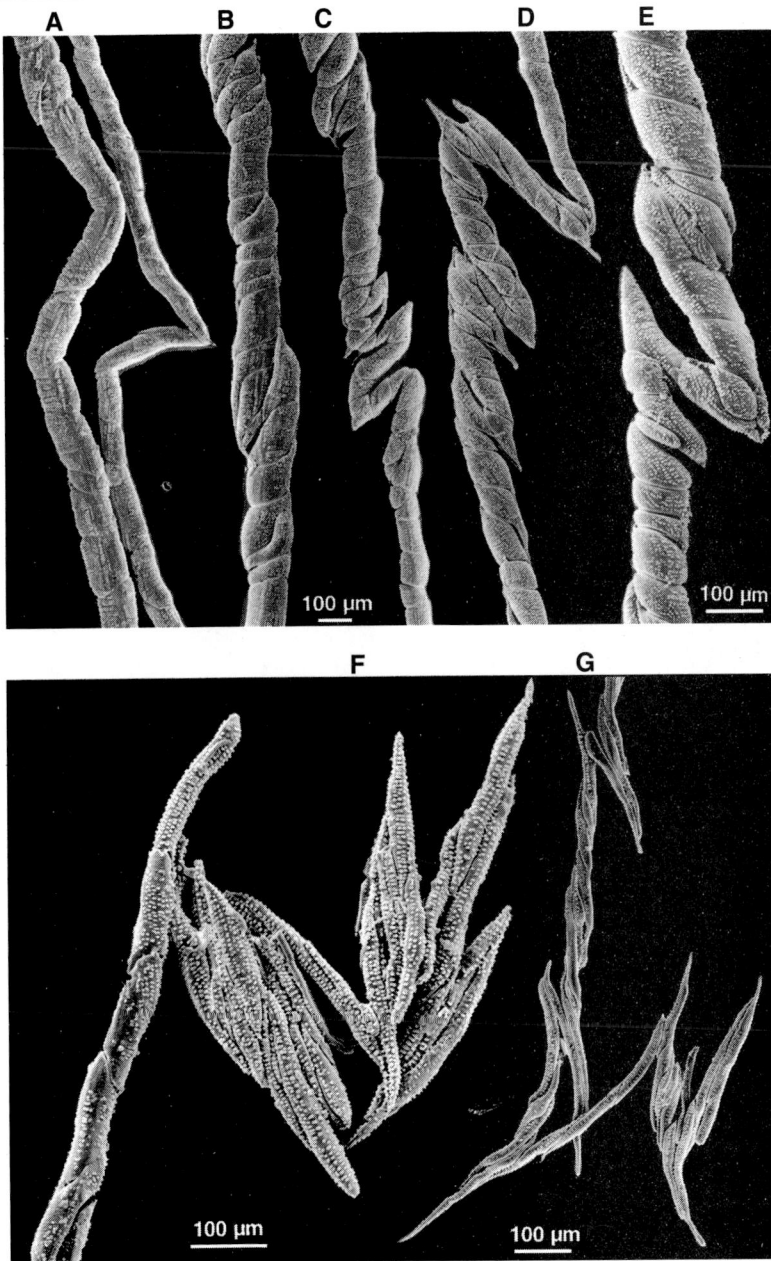

Plate 3.15. Zigzag vessel segments in spring wood (A to E) and in summer wood (F, G) in a branch of oak (*Quercus pedunculata*) (SEM photos). Casts A, B and C have conserved the appearance they have *in situ*. D and E are slightly stretched. F and G, which have the appearance of compact clusters *in situ*, are stretched here.

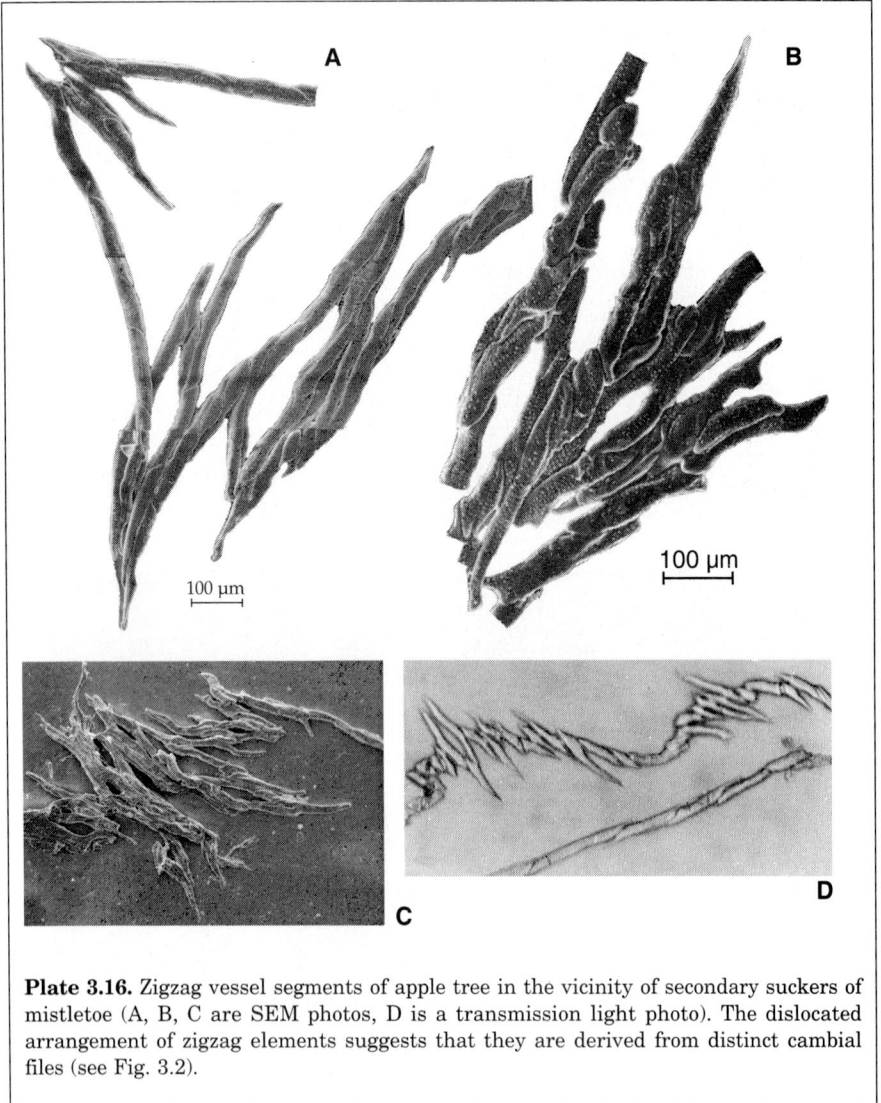

Plate 3.16. Zigzag vessel segments of apple tree in the vicinity of secondary suckers of mistletoe (A, B, C are SEM photos, D is a transmission light photo). The dislocated arrangement of zigzag elements suggests that they are derived from distinct cambial files (see Fig. 3.2).

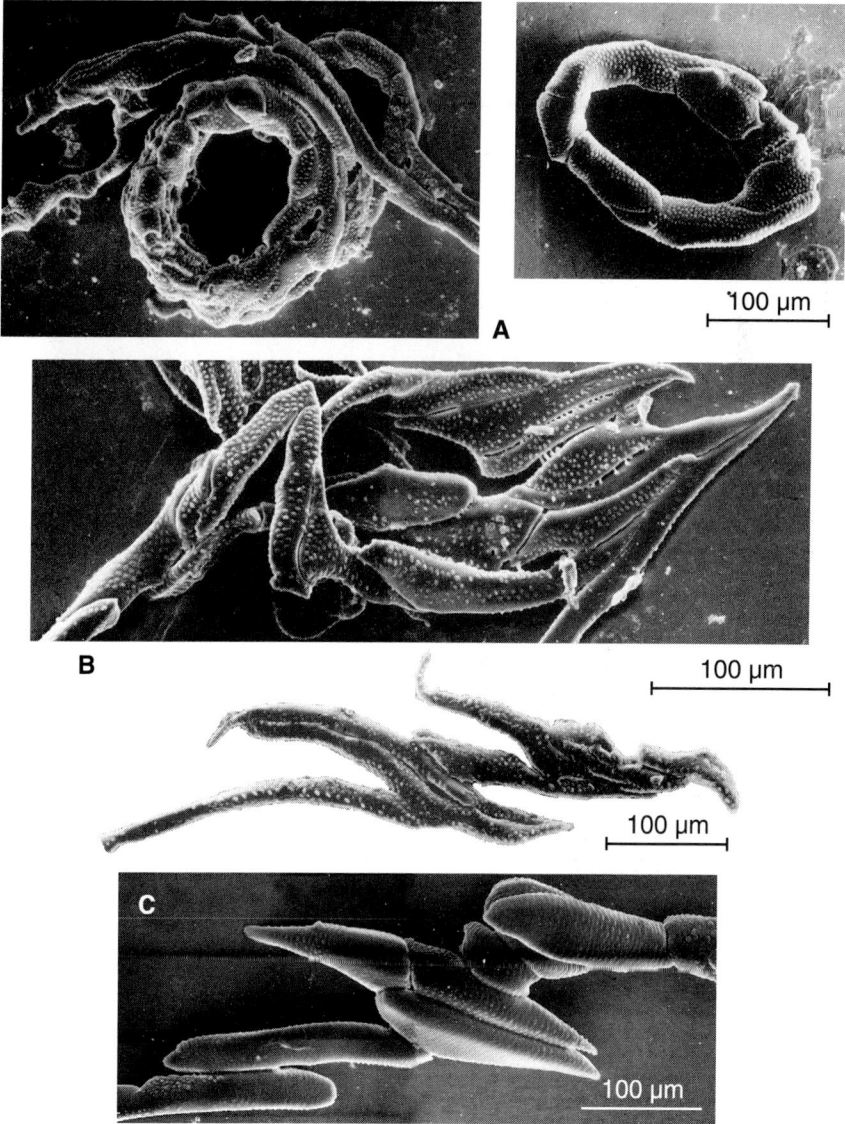

Plate 3.17. A. Zigzag segments (below), helical segments (left), annular segments (above right) in the newly formed secondary vessels in a gall induced on rose by *Agrobacterium tumefaciens* (crown gall) (see Fig. 3.3). B. Zigzag elements in a tomato stem node. C. Zigzag elements of metaxylem in maize (SEM photos).

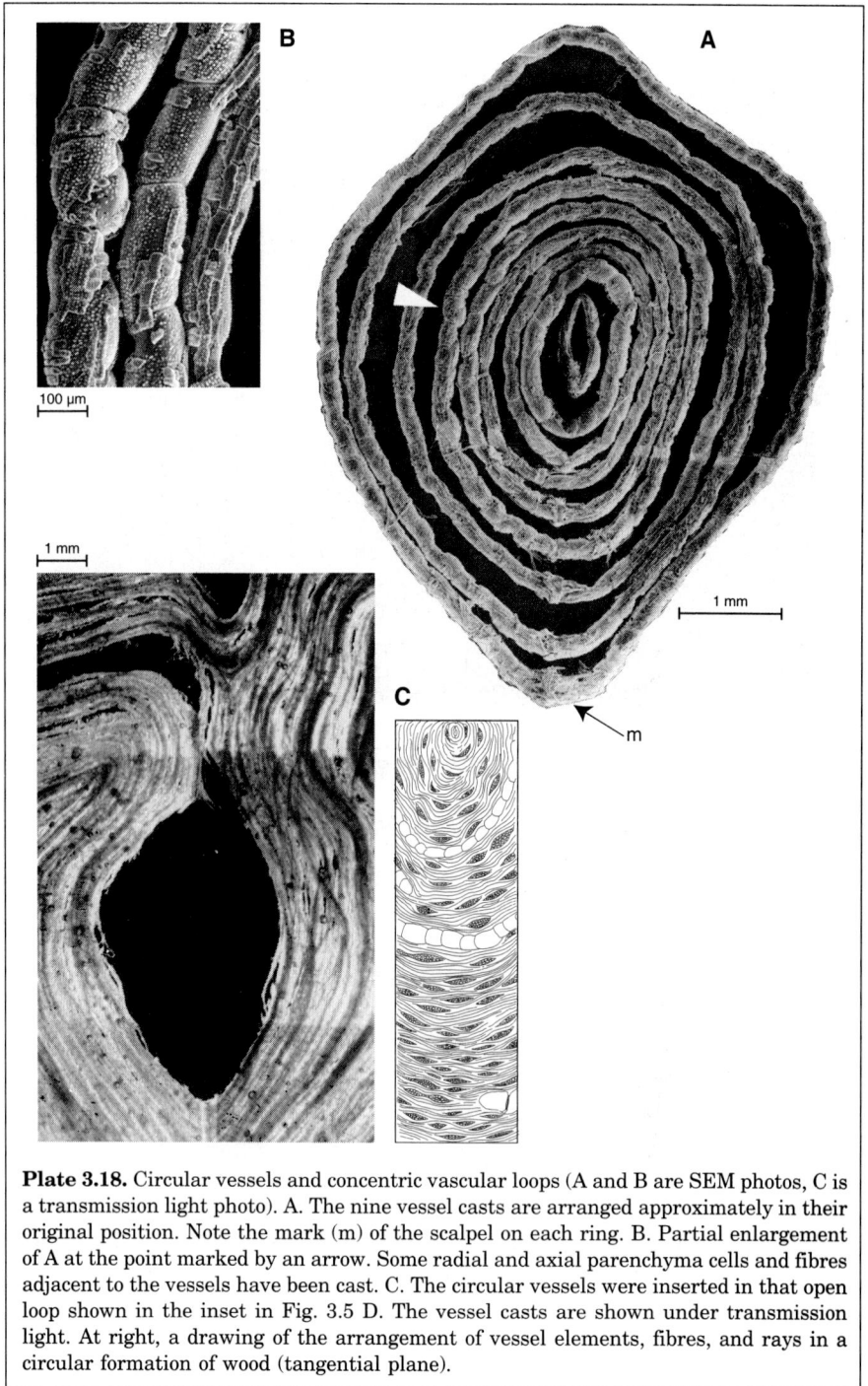

Plate 3.18. Circular vessels and concentric vascular loops (A and B are SEM photos, C is a transmission light photo). A. The nine vessel casts are arranged approximately in their original position. Note the mark (m) of the scalpel on each ring. B. Partial enlargement of A at the point marked by an arrow. Some radial and axial parenchyma cells and fibres adjacent to the vessels have been cast. C. The circular vessels were inserted in that open loop shown in the inset in Fig. 3.5 D. The vessel casts are shown under transmission light. At right, a drawing of the arrangement of vessel elements, fibres, and rays in a circular formation of wood (tangential plane).

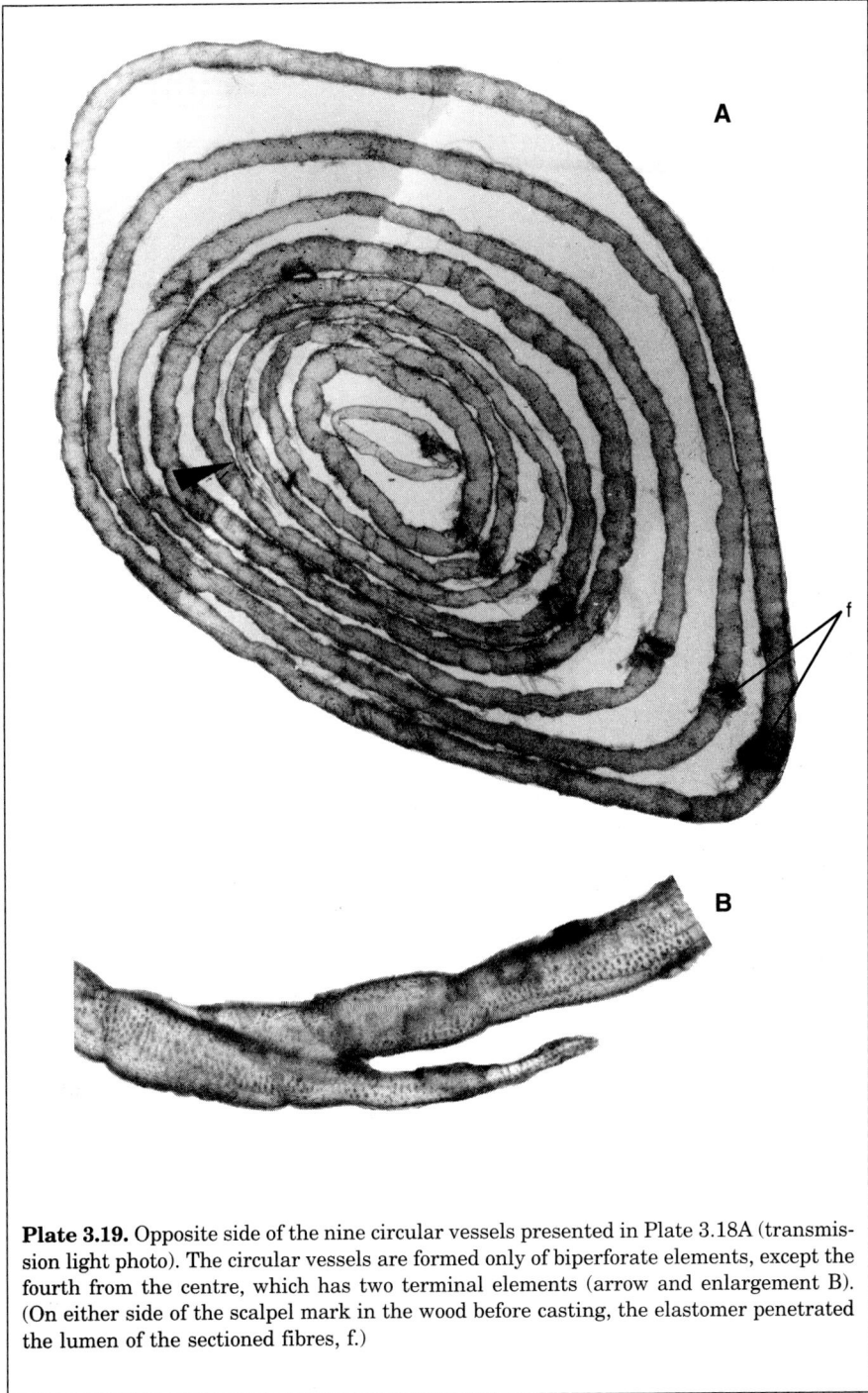

Plate 3.19. Opposite side of the nine circular vessels presented in Plate 3.18A (transmission light photo). The circular vessels are formed only of biperforate elements, except the fourth from the centre, which has two terminal elements (arrow and enlargement B). (On either side of the scalpel mark in the wood before casting, the elastomer penetrated the lumen of the sectioned fibres, f.)

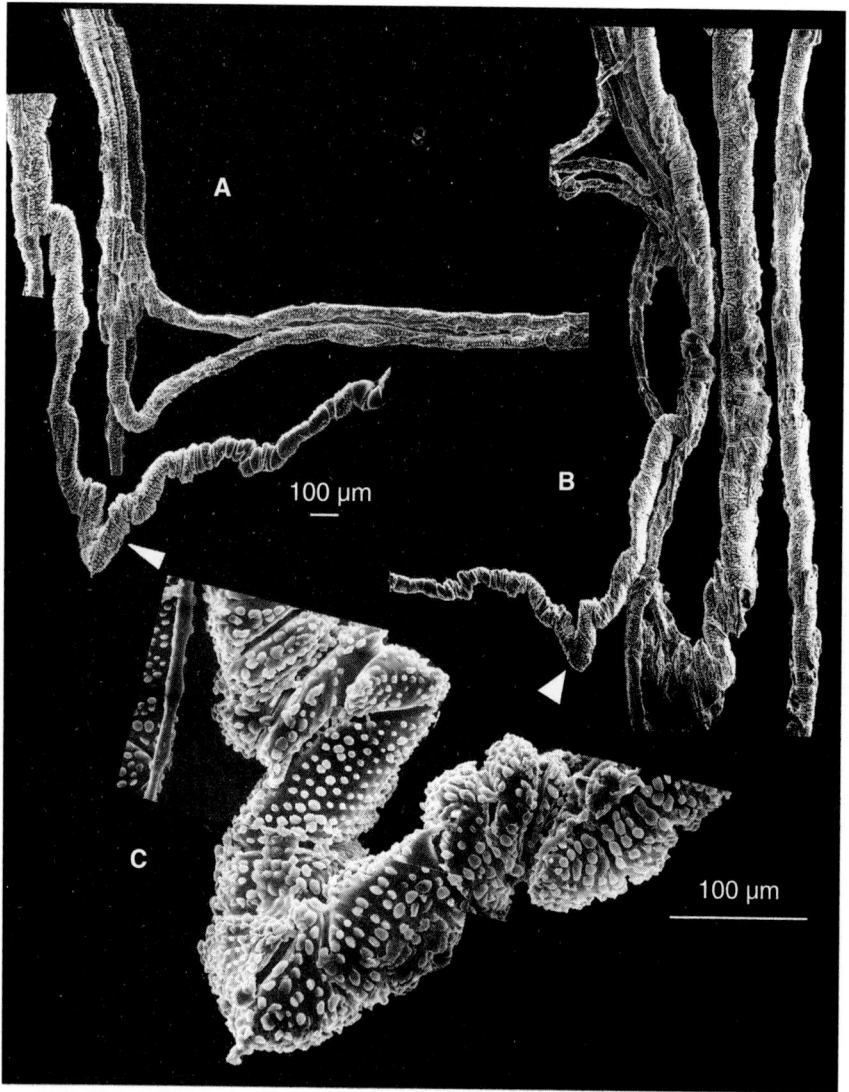

Plate 3.20. Secondary junction vessels between the stem and the adventitious roots of morning glory (*Ipomoea learii*). A, B. Views of a few vessel segments arranged as *in situ*. One of them is seen from both faces (arrows in A and B). C. Enlargement of U curve of this vessel (arrow in A). (SEM photos) (see Fig. 3.7).

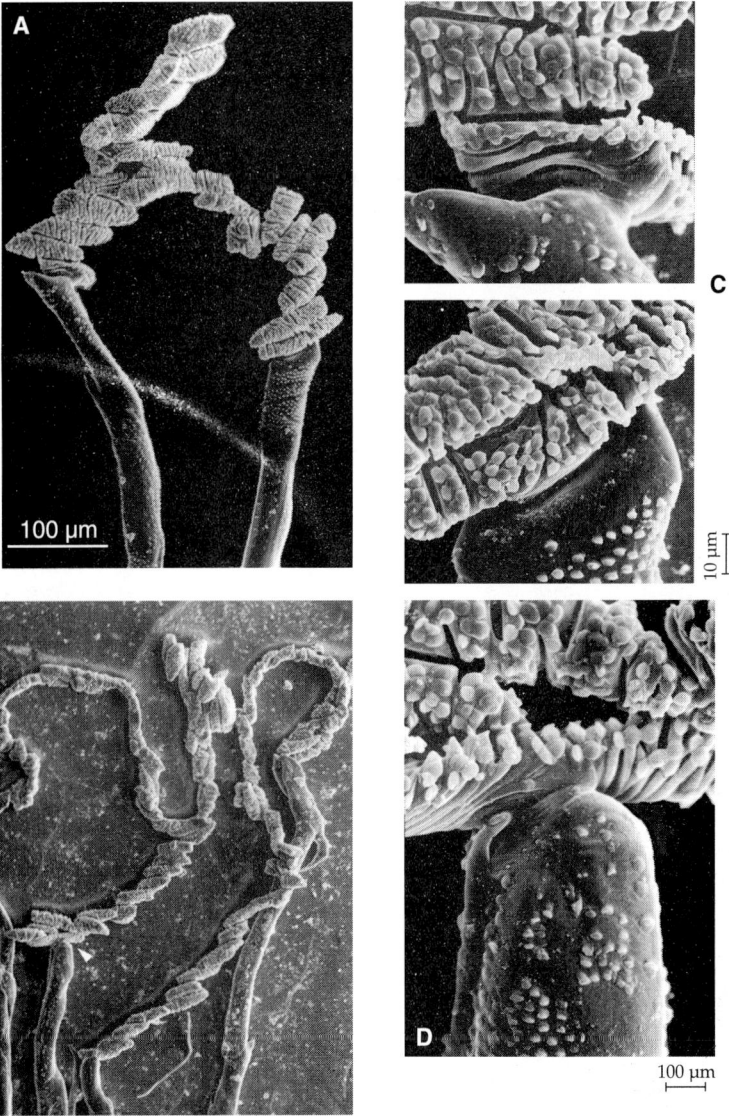

Plate 3.21. Connections between the host vessels and mistletoe pseudo-vessels (SEM photos). A. Two vessels of apple tree bridged by a chain of two pseudo-vessels linked by a common element. B. Other examples. Note the staircase-shaped juxtaposition of the pseudo-vessel elements. If the differentiation of pseudo-vessel elements is induced step by step from contact point with differentiating host vessels, it is conceivable that two differentiating chains converge randomly on the same cell and continue to extend into a single chain. C. Enlargements of junctions of A. D. Enlargement of a junction of B (arrow). The internal thickening of the wall of pseudo-vessels is characteristic. The bordered pits are numerous.

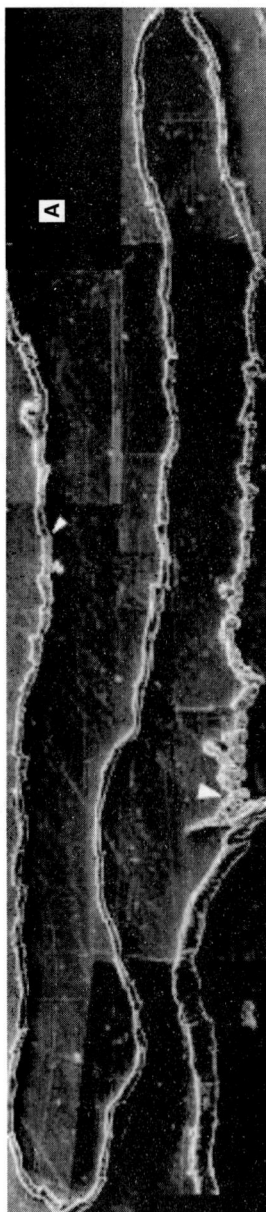

Plate 3.22. Vascular junction between a host wood vessel and a vessel of the parasite stem wood (SEM photos). A. Photographic mounting of the junction: the cast (length 12 mm) is folded for the purpose of the photograph. From bottom to top: vessel elements of the host up to the arrow, pseudo-vessels, then vessel elements of the parasite. B. Enlargement of one parasite element (arrowhead in part A of the figure). The quality of the cast in a long and thin duct is mediocre. C. Correct cast of vessel element of the parasite (cf. B). The drawing at left shows a segment of a parasite vessel.

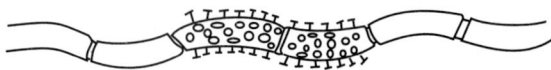

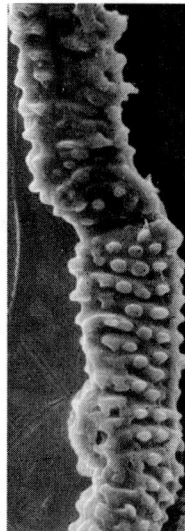

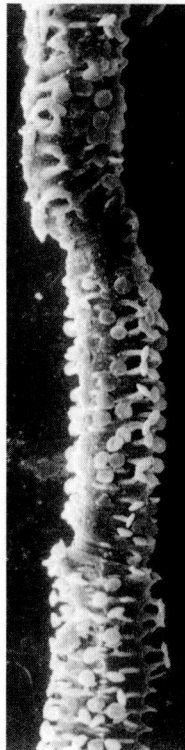

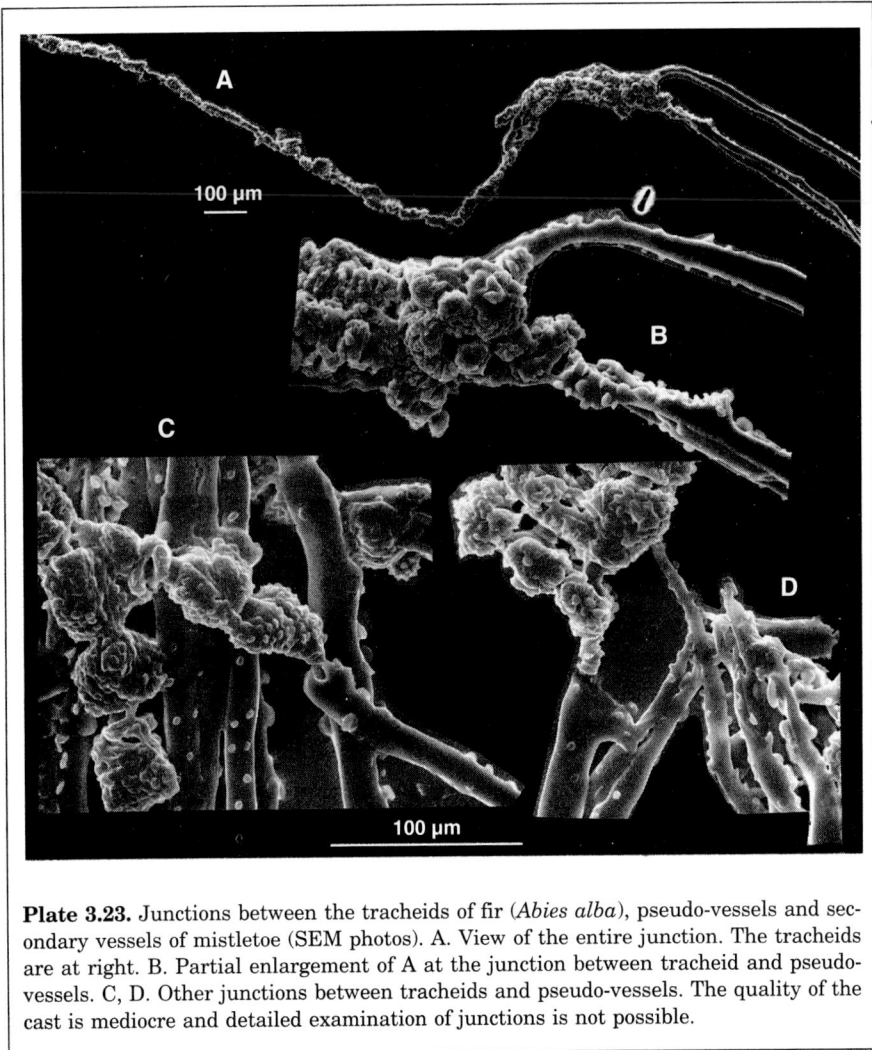

Plate 3.23. Junctions between the tracheids of fir (*Abies alba*), pseudo-vessels and secondary vessels of mistletoe (SEM photos). A. View of the entire junction. The tracheids are at right. B. Partial enlargement of A at the junction between tracheid and pseudo-vessels. C, D. Other junctions between tracheids and pseudo-vessels. The quality of the cast is mediocre and detailed examination of junctions is not possible.

The Xylem of Pteridophytes and Gymnosperms
(Intracellular casts)

ADVANTAGE OF CASTING OF TRACHEIDS

In the Pteridophytes and Gymnosperms, the conducting function of the xylem is not carried out by specialized longitudinal cell files, as it is with the vessels of Angiosperms. The tracheids that constitute most of the axial cells of the xylem participate collectively in this function. As a conducting unit, the tracheid is directly visible under the microscope because of its small dimensions unlike the vessel, and, therefore, the cast seems to offer little heuristic interest. The reality being slightly more complex, limited assays have nevertheless been made, motivated by the following questions. First, it is recognized that the tracheary elements of certain Pteridophytes could be perforated. Esau (1965) reports that elements with scalariform bordered pits of species belonging to the genera *Pteridium, Selaginella*, and *Equisetum* evolve into vessel elements by gradual perforation of the primary wall. At what point can the perforated tracheary elements constitute continuous lines analogous to vessels? Second, the pits of the tracheids of Gymnosperms are partly perforated in the torus margo, where the primary wall is reduced to some cellulosic fibrils. Does their permeability make possible the injection of the casting agent in non-sectioned tracheids? Some answers have been found to these questions.

THE XYLEM OF PTERIDOPHYTES

The present-day Pteridophytes are grouped in orders of widely varying size: only 30 species in the Equisetales, 200 in the Lycopodiales, 800 in the Selaginellales, and around 10,000 in the Filicales (ferns). All these lineages, the ancestors of which were widespread 300 to 400 million years ago, are endangered (Emberger, 1960). The observations discussed here pertain to five species.

The two *Selaginella* examined, from the coastal regions of Guyana, are small herbaceous plants with spread-out branches (Plate 4.1). Their identification (*S. sulcata* and *S. inaequifolia*) from the available data is uncertain (Mori, 1997). Referred to here as S_1 and S_2, they differ mainly in the number of steles in their stems. The lycopodium (*Lycopodiella cernua*), from the same region, is an erect herbaceous plant (Plate 4.3). The fertile axes of horsetail (*Equisetum hyemale*) and the fronds of the bracken fern (*Pteridium aquilinum*) were collected in France (Plates 4.4 and 4.5).

These Pteridophytes do not form secondary tissues. The primary xylem and the phloem differentiate from the procambium. When they are formed, they constitute one or several steles, isolated from the cortex by the two concentric cell layers of the pericycle and the suberized endodermis. In a cross-section of stele, the tracheids appear in adjacent files of increasing diameter, those of the protoxylem smaller at the periphery, those of the metaxylem larger at the centre (Plates 4.1 to 4.5). The casts were done on air-dried segments of stem and rachis (fern).

• *Selaginella* and *Lycopodium*

The elastomer penetrates the sectioned tracheids, then penetrates several successive rows of entire tracheids, from which a complete cast can be obtained (Plate 4.1D). Primary walls are thus perforated at the level of certain pits. An example is the cast of tracheids of S_2 linked by pit pairs soldered by the elastomer at the level of their pit chambers (Plate 4.2B). The tracheids are tapered at both ends and of uniform and regular morphology in the three species; a small number of them are forked (Plate 4.3D). The secondary walls are of helical and scalariform types.

The lateral perforations of the primary wall favour lateral water transport, as is the case between the adjacent tracheids of Gymnosperms (see below), but there is no formation of long longitudinal files of pitted tracheids analogous to vessels.

• Horsetail

The internodes of the stem grow from an intercalary basal meristem like the internodes of grass culms. Their relatively soft base is sheathed in the rigid leaf whorl of the underlying internode (Plate 4.4B).

When a nodal segment is cast, the elastomer penetrates only the $\alpha\beta$ part of the underlying internode (Plate 4.4D). The cast of the metaxylem is made of fragile and fine filaments, with circular pits of large diameter, set in regular files (Plate 4.4E, F). It is difficult to locate the tips of the tracheids (Plate 4.4F).

The elastomer invades the carinal canal, which contains the remains of the protoxylem, i.e., the rings of the secondary wall. The rings are embedded in the cast and visible under transmission light microscope. In the nodal zone, indicated by point β, the canal is wider. There are finely pitted cells in the cast of this part.

• Bracken fern

In the rachis are isolated steles that divide, following ramifications of orders 1, 2, and 3 of the frond. The elastomer penetrates deep into the xylem. Casts of entire tracheids, tapered and of varying lengths, and of entire vessels formed from a small number of elements are obtained (Plate 4.5C). The cells cast have regular scalariform pits on their planes of mutual contact and scalariform perforations between the vessel elements. The cast of entire cells implies that some lateral pits are perforated at least partly. The casting technique allows us to explore in detail and conveniently the pathways of water circulation through natural perforations in the mass of tracheids and of vessels in the Pteridophytes.

THE XYLEM OF GYMNOSPERMS

The 650 present-day species of conifers constitute by far the largest group of Gymnosperms. They are distributed in the orders Pinales (300 species), Araucariales (40 species), Podocarpales (140), Cupressales (160 species), and Taxales (10 species) (Emberger, 1960). In addition to the conifers we should mention the ginkgo (*Ginkgo biloba*), a unique survivor of a family that developed during the Jurassic. This species, in which the ovules become detached after pollination (see Table 1), is included with the Gymnosperms, which it precedes from an evolutionary point of view, on the basis of several anatomical characters.

In these perennial and arborescent species, the cambium forms a wood with tracheids. The observations of casts of air-dried wood are limited to five species: *Sequoiadendron gigantea* and *Cupressus sempervirens* (Cupressales), *Taxus baccata* (Taxales), *Pinus mugo* (Pinales), and *Ginkgo biloba* (Ginkgoales) (Plates 4.6, 4.7, 4.8, 5.1).

The elastomer easily passes through the pits with torus of the tracheids, mainly the lateral pits, over several tens of successive tracheid rows, in the tangential plane. In the cast of a block of wood, the lumen casts of the tracheids stick to one another by their soldered lateral pit pairs and separate into concentric sheets.

The cast of soldered pit pairs has the pedagogical merit of materializing the volume of the sap circulating through the tracheids, as well as of showing one of the anatomical characters common to ginkgo and the conifers. It could no doubt also illustrate the stop valve function served *in vivo* by the torus in case of accidental local embolisms.

More interesting could be the possibility of analysing, conveniently and on as many tracheids as needed, the stochastic determinism of the direction and number of gyres of tertiary thickening that form the last layer of the secondary wall. Methodologically, it should be advisable to compare the practical advantage of casting with the direct examination of wood under SEM and with the technique of maceration.

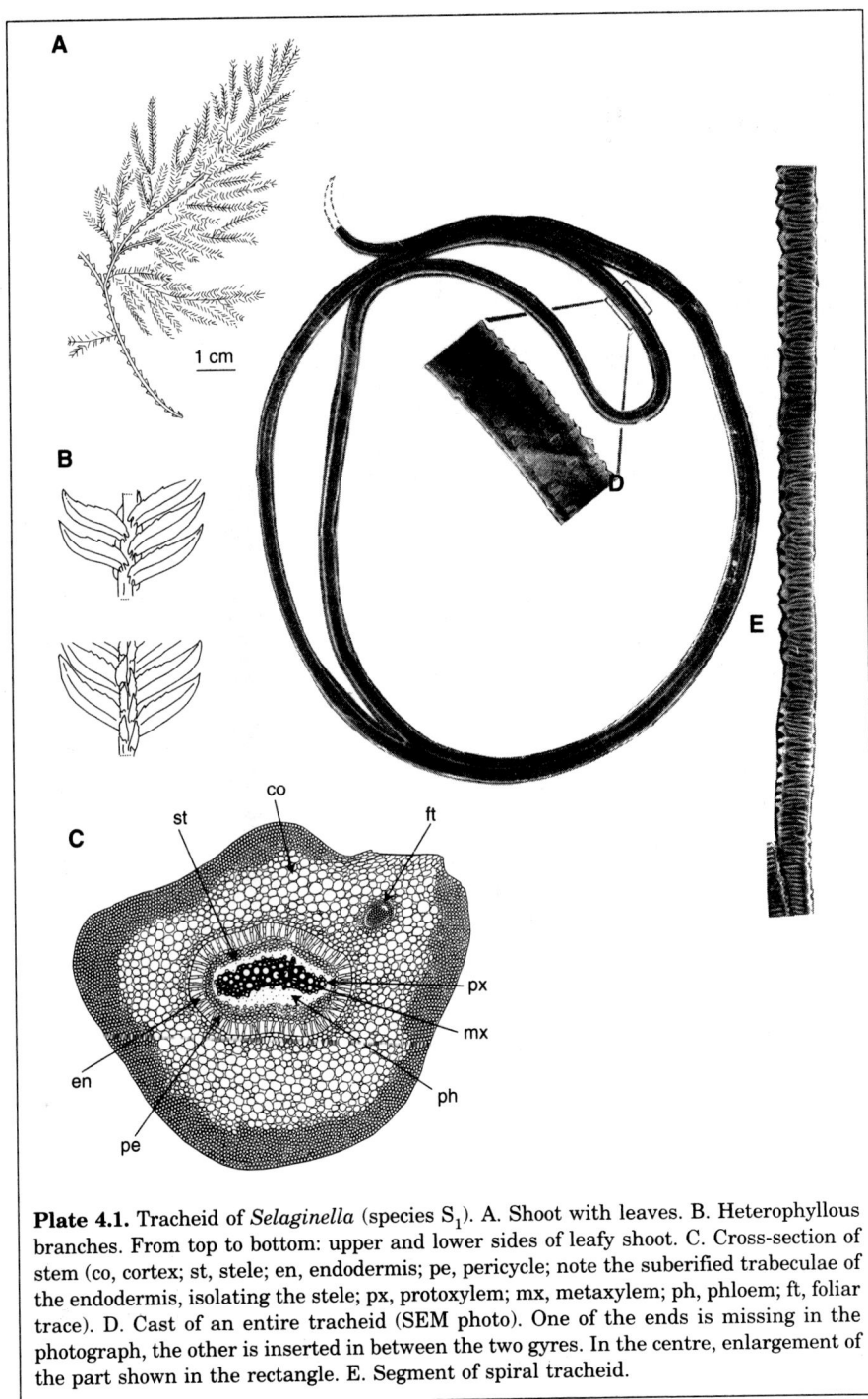

Plate 4.1. Tracheid of *Selaginella* (species S₁). A. Shoot with leaves. B. Heterophyllous branches. From top to bottom: upper and lower sides of leafy shoot. C. Cross-section of stem (co, cortex; st, stele; en, endodermis; pe, pericycle; note the suberified trabeculae of the endodermis, isolating the stele; px, protoxylem; mx, metaxylem; ph, phloem; ft, foliar trace). D. Cast of an entire tracheid (SEM photo). One of the ends is missing in the photograph, the other is inserted in between the two gyres. In the centre, enlargement of the part shown in the rectangle. E. Segment of spiral tracheid.

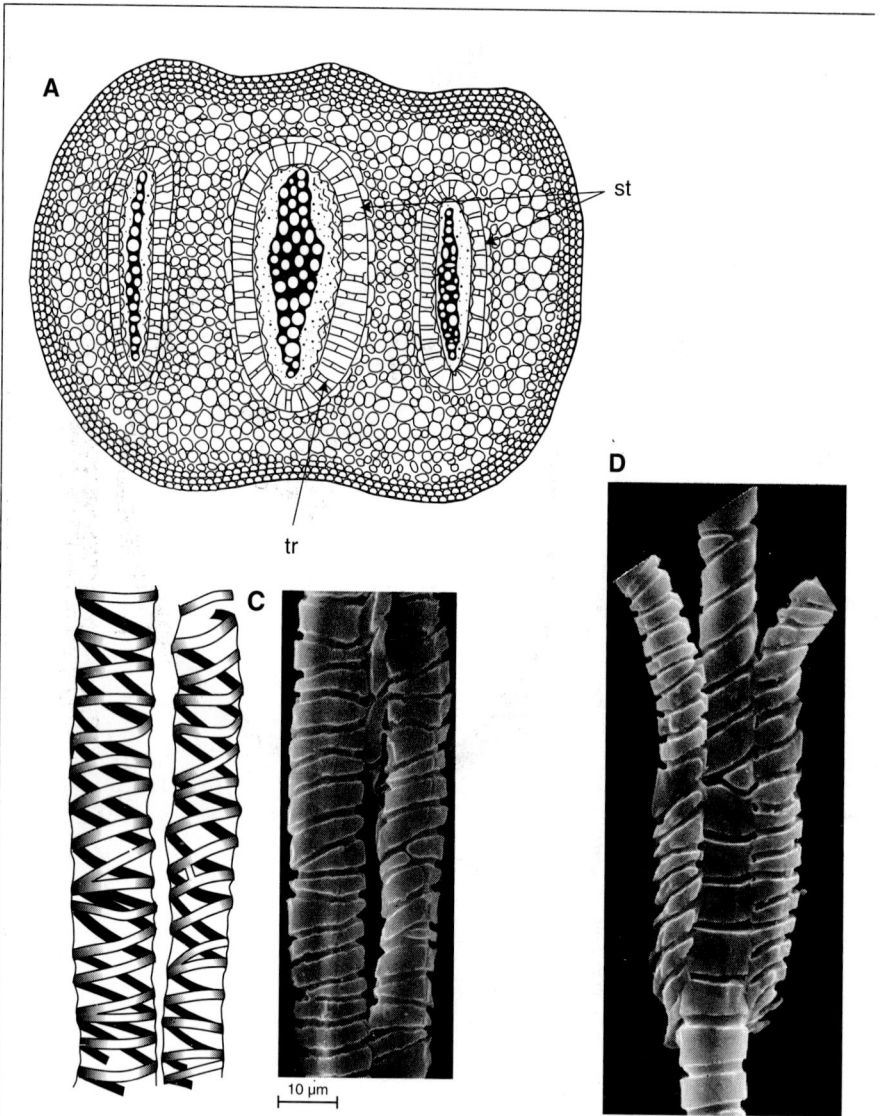

Plate 4.2. Tracheids of *Selaginella* (species S_2). A. Cross-section of stem containing three steles (st) isolated by suberized trabeculae (tr). B. (see p. 91) Segments of tracheids with scalariform pits. The pairs of bordered pits visible at right are soldered by the elastomer. In the middle, towards the background, end of the lumen of a tracheid, linked to its two neighbours by pits. C. Segments of two tracheids, with a spiral fissured secondary wall. At left, a drawing of the secondary wall. D. Ends of two tracheids, adjacent to a third, partly spread out for the photograph (SEM photos).

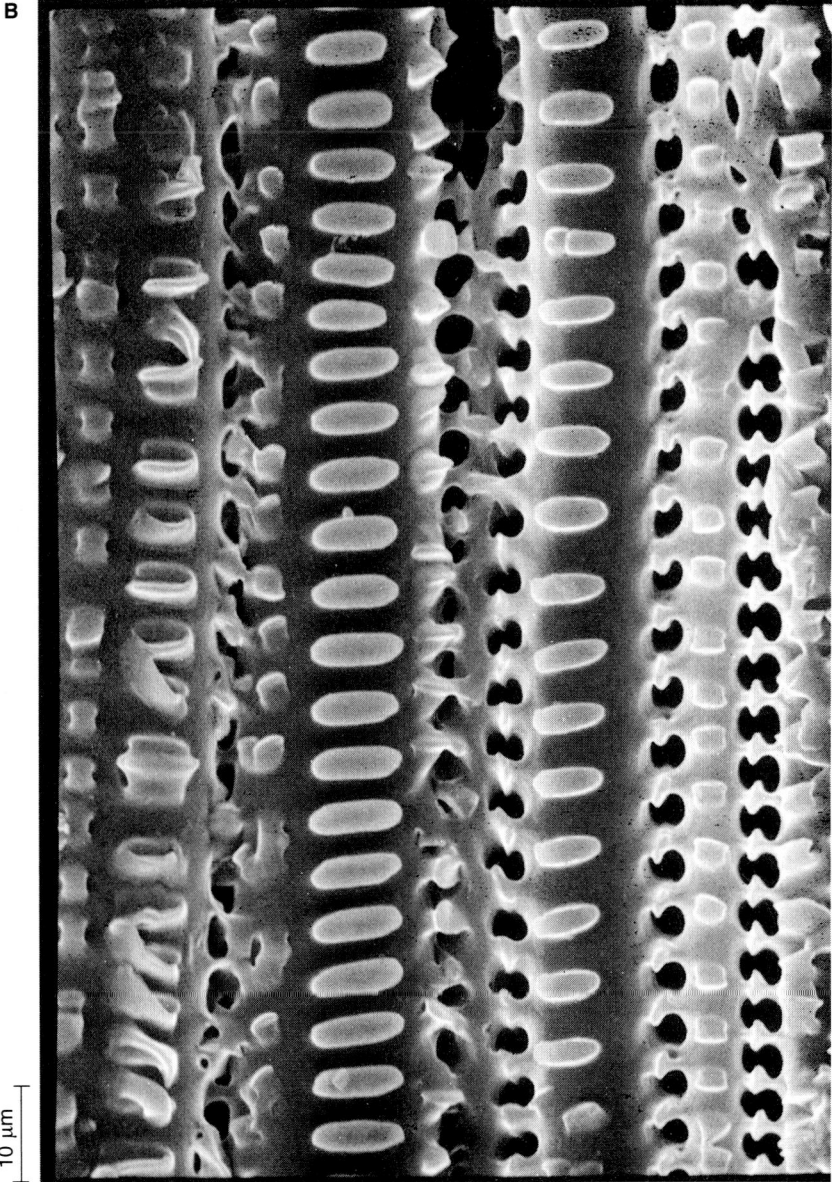

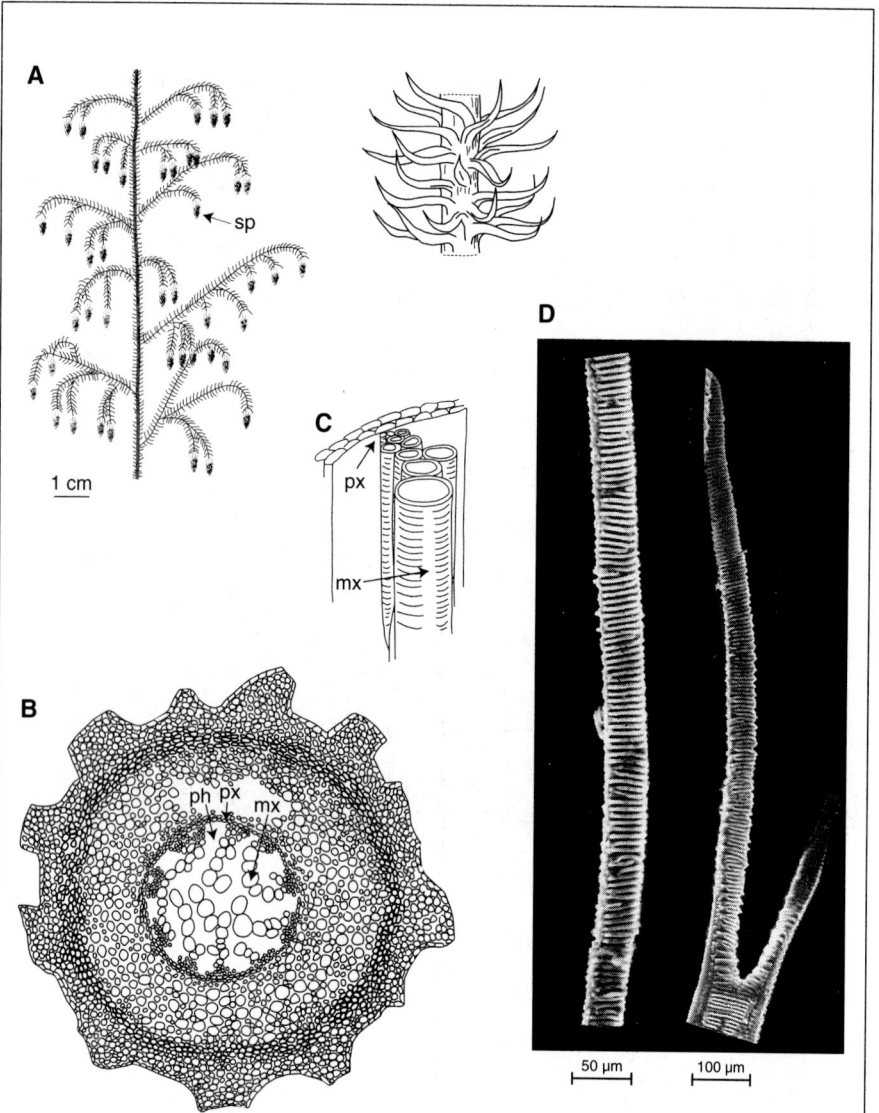

Plate 4.3. Tracheids of *Lycopodiella*. A. Branched stem with needle-like leaves. Hanging sporangiferous spike (sp) at the tip of each branch. B. Cross-section of stem: several protoxylem poles. C. Drawing of tracheids in the stele. D. Segments of tracheids with spiral walls. At right, forked end with scalariform pits at the base of the fork (SEM photos) (px, protoxylem; mx, metaxylem).

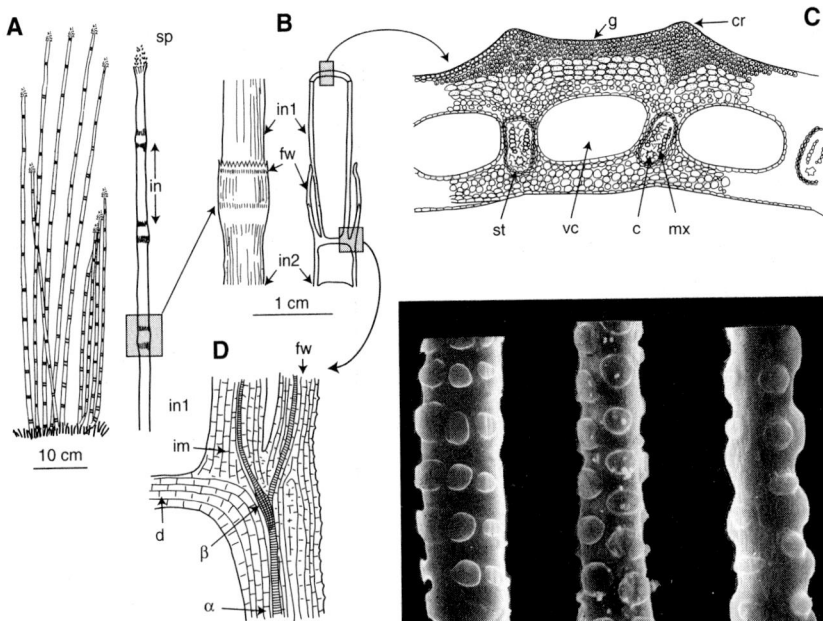

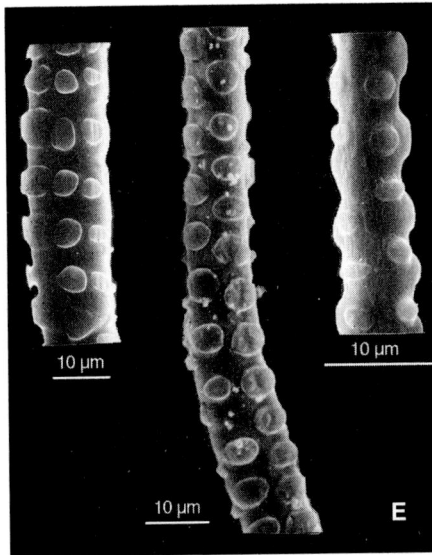

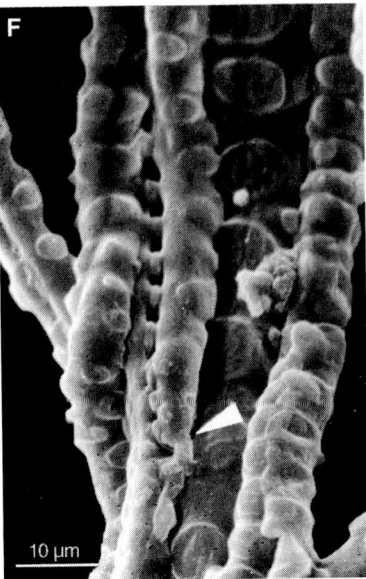

Plate 4.4. Tracheids of horsetail (*Equisetum hyemale*) (SEM photos). A. Fertile stems terminating in a sporangiferous spike (sp). The stems are made up of internodes (in). B. Nodal segment. The insets are shown enlarged in C and D. (fw, foliar whorl; in1, upper internode; in2, lower internode.) C. Cross-section of the internode. Each internode contains up to 50 steles (st) and an equal number of vallecular canals (vc), located in the grooves (g) between two crests (cr) of the epidermis. Once the protoxylem is non-functional, it gives way to a carinal canal (c). (mx, metaxylem.) D. Radial section of the node at the junction between in1 and in2 (im, intercalary meristem zone; d, diaphragm; αβ, see text). E. Segments of tracheids with circular pits: The bordered character is not visible. F. The same, in a defective cast. At centre, lower end of a tracheid (arrow) and junctions of pit pairs. Note the small diameter of the tracheids.

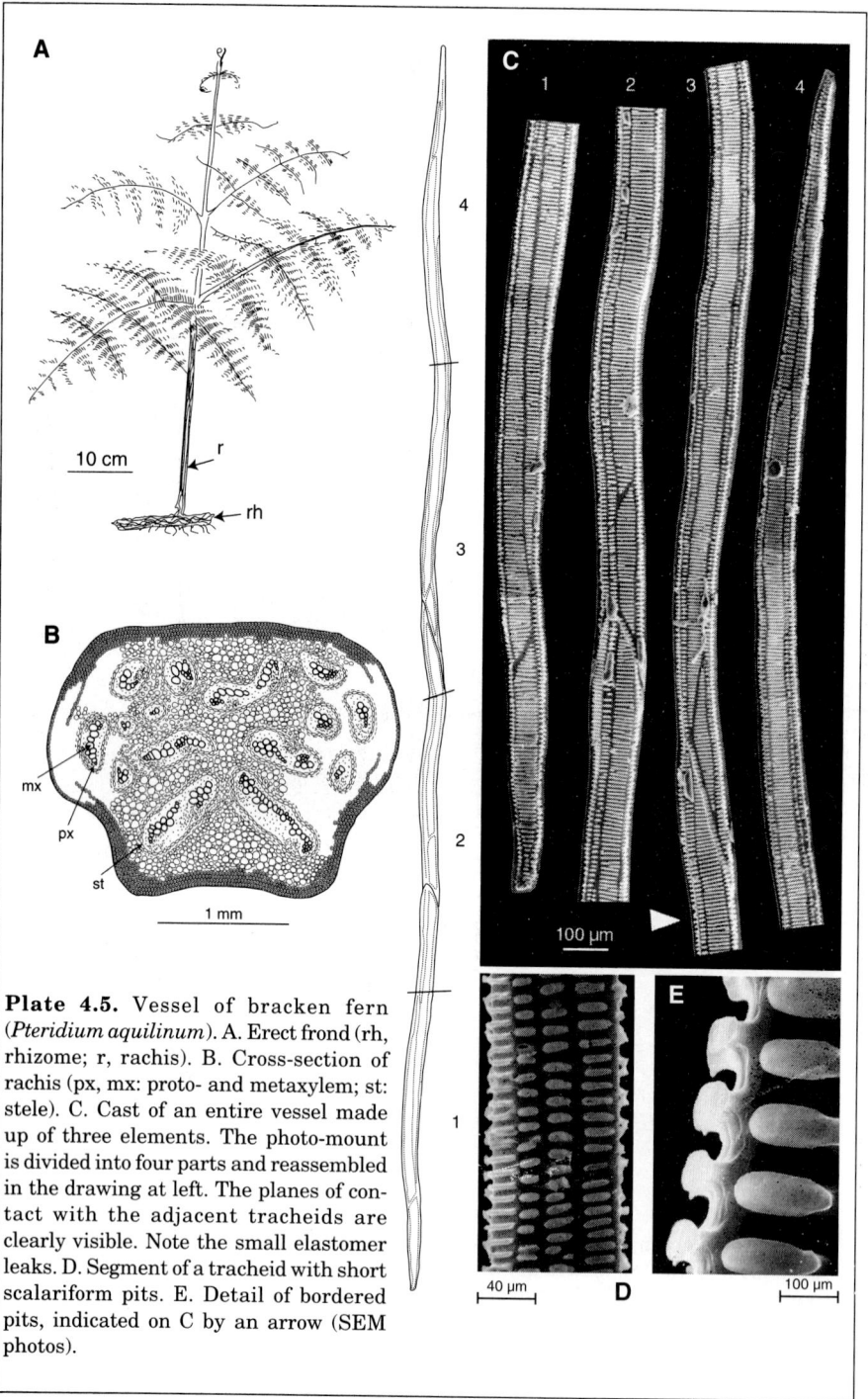

Plate 4.5. Vessel of bracken fern (*Pteridium aquilinum*). A. Erect frond (rh, rhizome; r, rachis). B. Cross-section of rachis (px, mx: proto- and metaxylem; st: stele). C. Cast of an entire vessel made up of three elements. The photo-mount is divided into four parts and reassembled in the drawing at left. The planes of contact with the adjacent tracheids are clearly visible. Note the small elastomer leaks. D. Segment of a tracheid with short scalariform pits. E. Detail of bordered pits, indicated on C by an arrow (SEM photos).

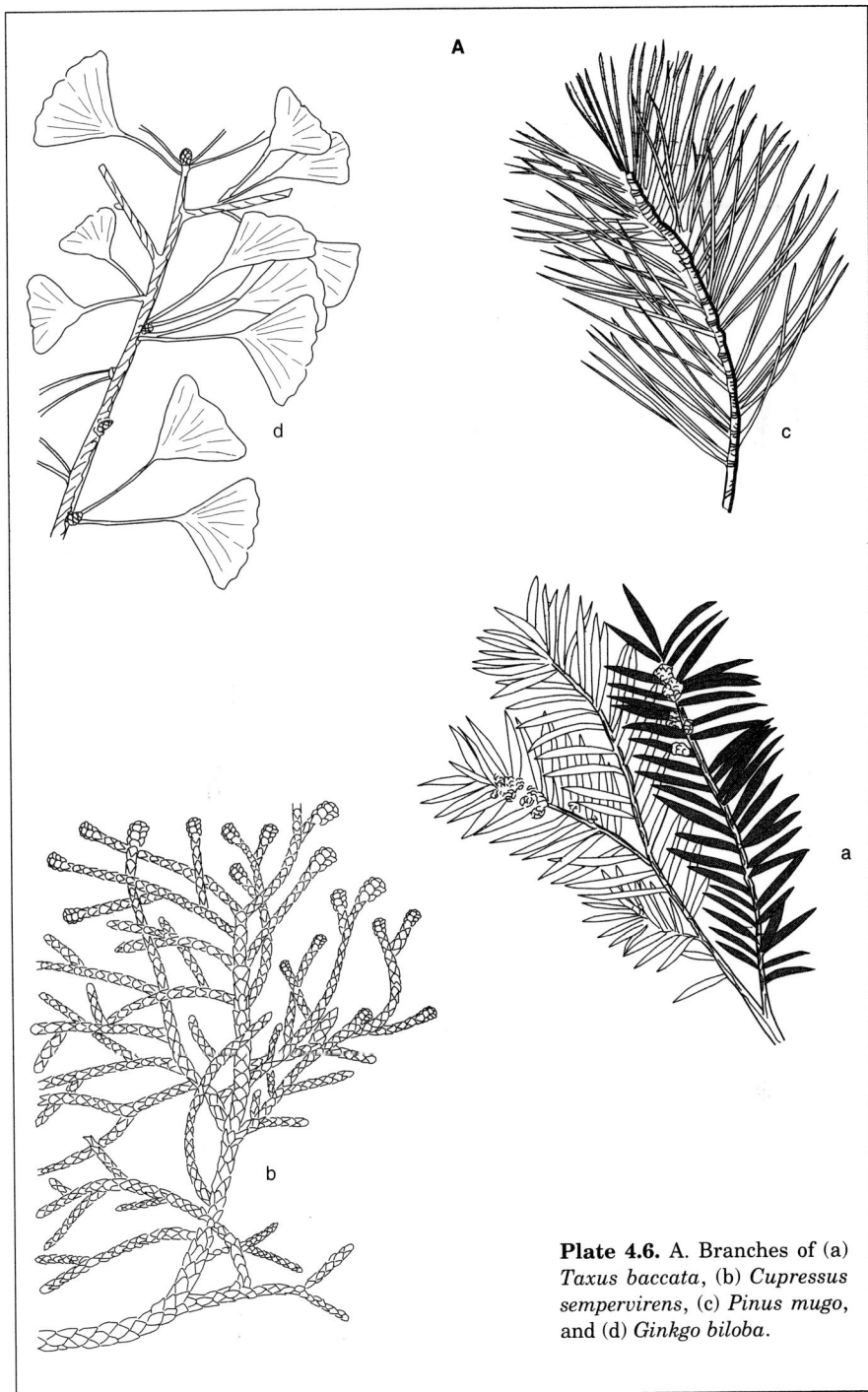

Plate 4.6. A. Branches of (a) *Taxus baccata*, (b) *Cupressus sempervirens*, (c) *Pinus mugo*, and (d) *Ginkgo biloba*.

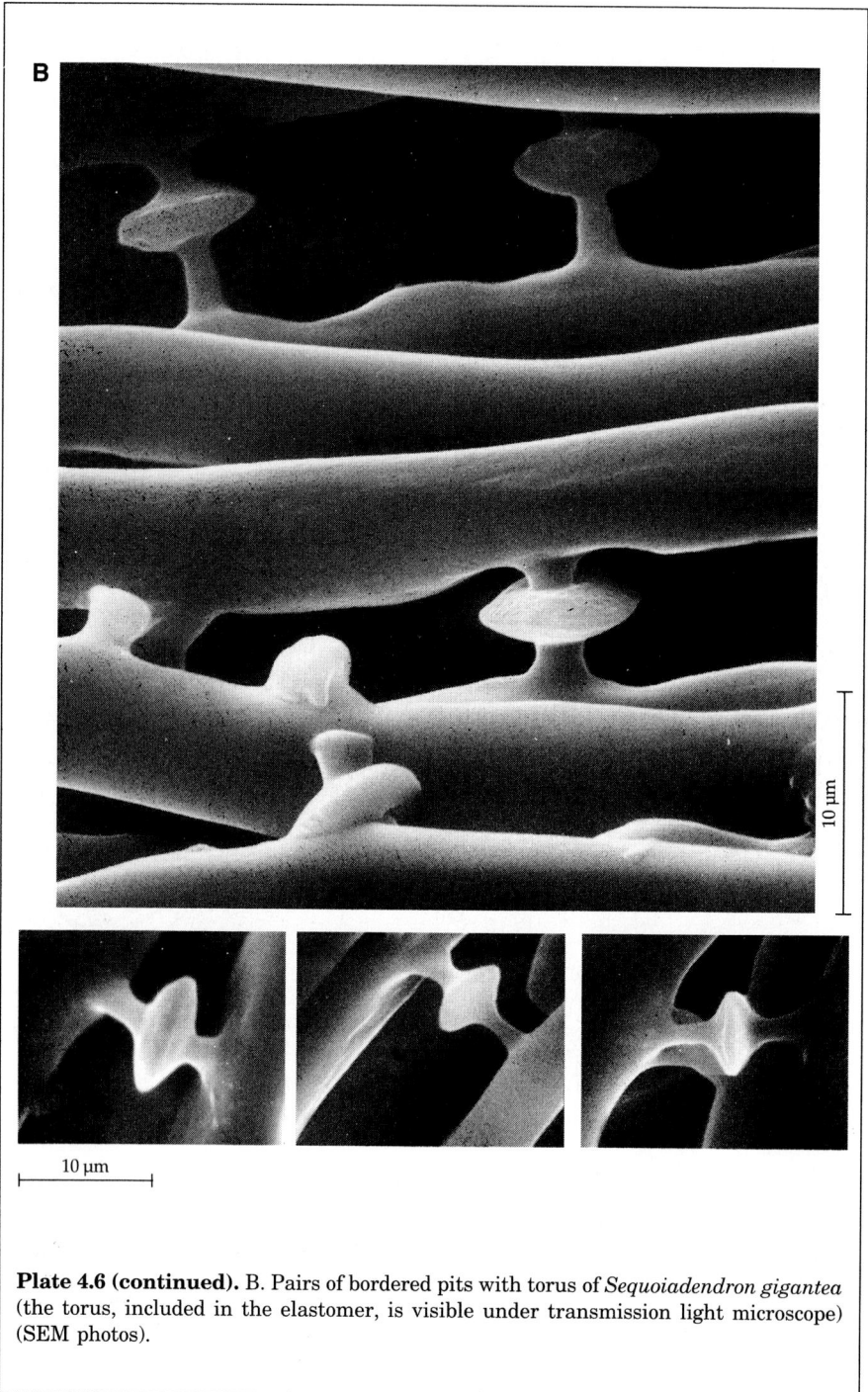

Plate 4.6 (continued). B. Pairs of bordered pits with torus of *Sequoiadendron gigantea* (the torus, included in the elastomer, is visible under transmission light microscope) (SEM photos).

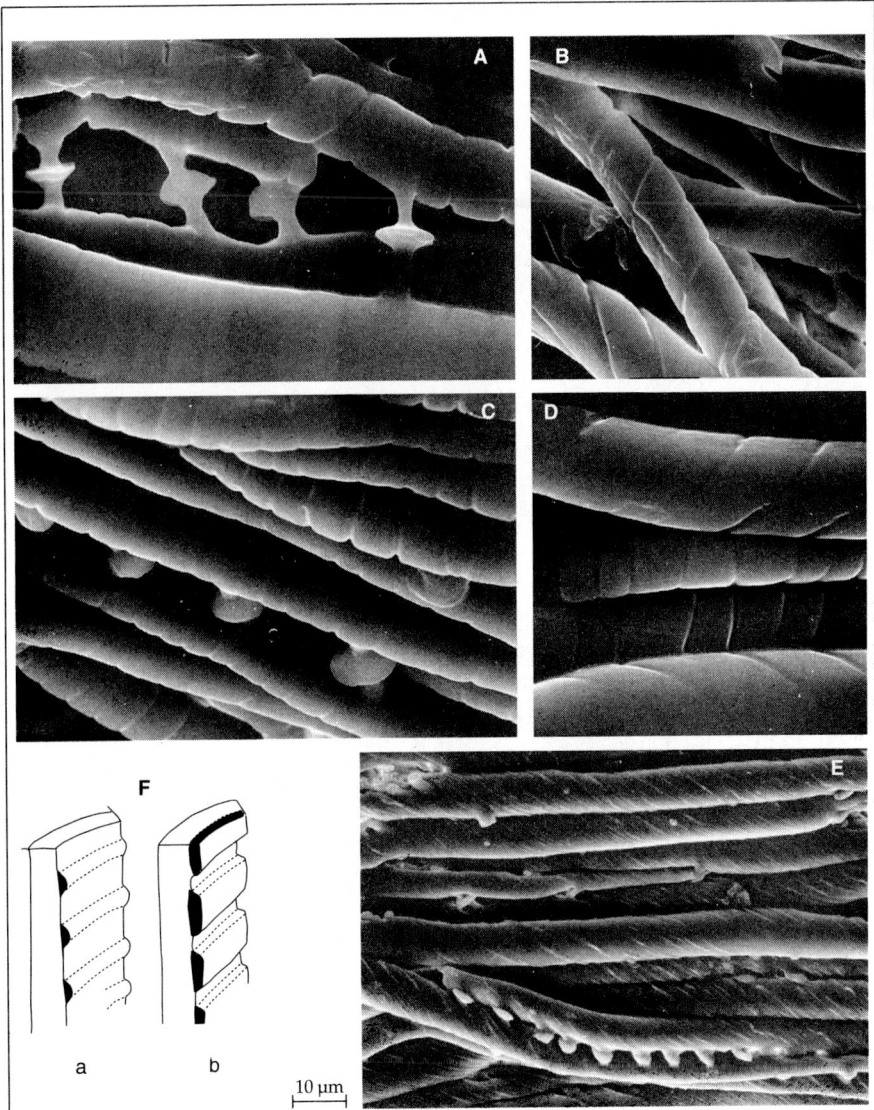

Plate 4.7. Tertiary wall thickenings on the inner surface of the tracheids of yew *Taxus baccata* (A to D) and cypress *Cupressus sempervirens* (E). A to D. In *T. baccata*, the narrow thickenings appear as grooves at the cast surface. They are right-hand spirals (D lower part), left-hand spirals (D upper part), simple spirals (D centre), or double spirals (D upper and lower parts). (See section 1.) E. In *C. sempervirens*, on the contrary, the thickenings are larger and separated by grooves that appear as relief in the casts. All the gyres are right-hand in the branch segment examined (SEM photos). F. Drawing of thickenings seen in section: (a) *Taxus* type, (b) *Cupressus* type.

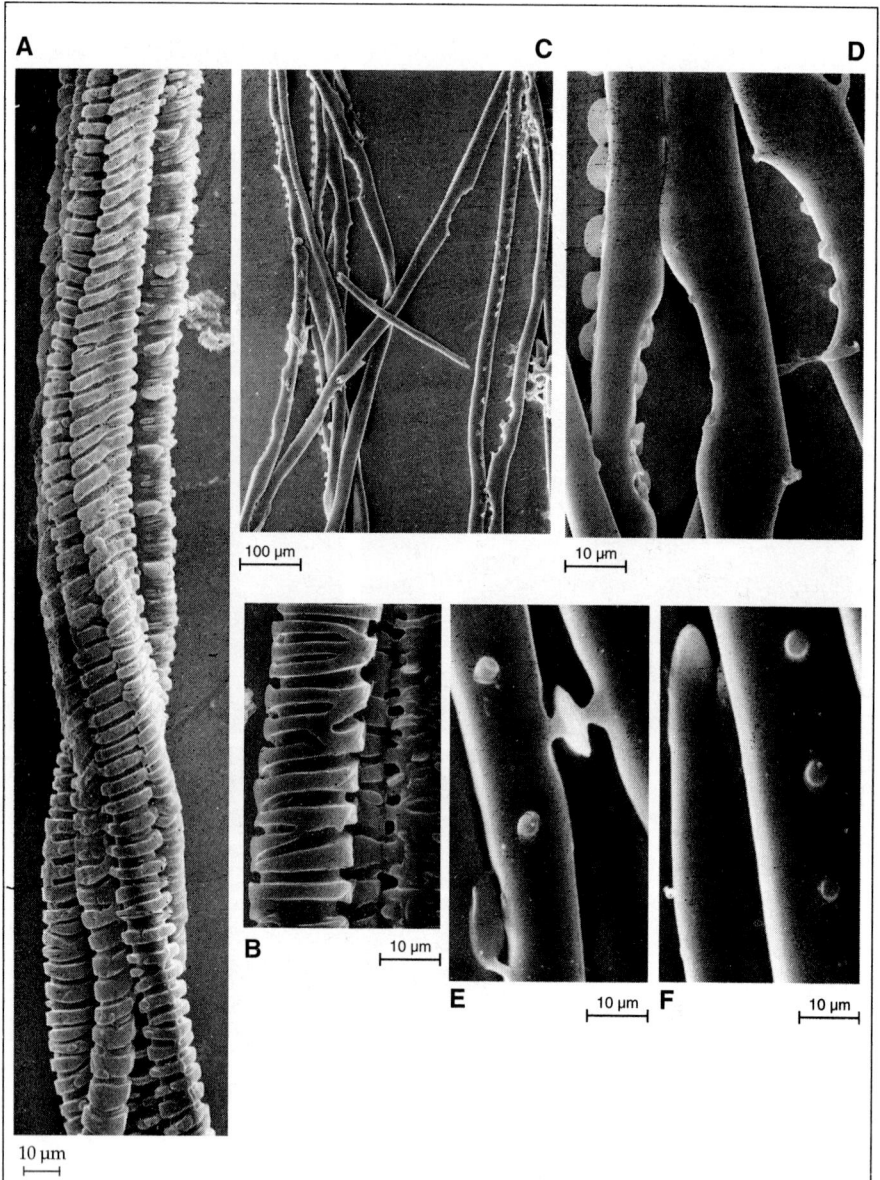

Plate 4.8. Tracheids of *Ginkgo biloba*. A. Segments of tracheids of primary xylem, with spiral-annular walls. The tracheid at right also seems to present equidistant pits. B. Segments of tracheids with spiral walls with incomplete gyres. C. Tracheids of the secondary xylem (wood) with bordered pits with torus. The tracheids adjacent to rays have a lateral indentation. D. Detail of C: lateral indentations with tracheid-ray pits. E. Joined pair of bordered pits with torus, identical to that of conifers. F. End of tracheid. The casts of pits break by traction at the pit canal level (SEM photos).

Intercellular Spaces and Canals

(Intercellular casts)

Phloem and Parenchyma

(Intracellular casts)

ADVANTAGE OF CASTING

This last section of Part I gives an overview of other possible applications of histological casting. In fact, the technique is not limited to the exploration of xylem conducting elements. The cutting of air-dried tissues inevitably gives the casting agent access to every intercellular and intracellular space and canal, as well as to certain files of parenchyma cells and, under particular conditions, to the phloem.

Let us briefly examine the advantage of these different types of casts.

CASTING SPACES AND CANALS

The canals and spaces are intercellular spaces of accumulation, storage, and circulation of liquid substances produced by the bordering cells (resin, gum, and latex canals) or of gas of composition similar to that of air (air spaces). The examination of casts of the three-dimensional extension of a network of air spaces or ducts could be of anatomical or physiological interest.

One example is that of resin canals of Pinaceae. The secretory cells are differentiated in the parenchyma of the cortex, of the pith, and of the wood. The canals in which the resin accumulates form distinct networks in each of the tissues. Physical and parasitic traumas affect resin secretion and circulation. The injection of a casting agent after the extraction of resin (see Part II, pp. 116–117) reveals the spatial structure of these networks (Plate 5.1).

As a second example, the highly reticulated intercellular spaces, full of air, of the hypocotyls of *Vigna radiata* have been used as a pathway for access to the surface of cells of the cortex and of the central cylinder in studies on primary wall and plasma membrane permeability (Prat et al., 1997). Previously, the network of air spaces was explored by casting on fresh tissue (Plate 5.2).

The volume of the intercellular space that extends into the mesophyll of leaves and petals of many species can easily be determined by the weight of its cast, measured on fresh tissue. Injection into the leaf, via the stomata, fixes the image of their degree of opening at various points of the organ (Plates 5.3 and 5.4).

INTRACELLULAR CASTING IN PARENCHYMA

The casting of parenchyma cells is generally confined to the immediate vicinity of cross-sections. The penetration of the elastomer into whole cells depends on still undetermined factors that affect the permeability of the primary wall or cause it to tear (see Part II, p. 117). A preliminary freeze-drying often favours the casting of aggregates of whole cells in the medullary parenchyma and also, via the adjacent vessels, in the radial parenchyma (Plates 5.5 and 5.6).

Casts of aggregates of whole cells could offer, for cellular morphometric analysis, a simple alternative to confocal microscopy by laser scanning (Gray et al., 1999).

CASTING IN THE SIEVE TUBES OF PHLOEM

The organization of the sieve tube network of the phloem is in itself an important subject of study in which our anatomical understanding is still superficial (J.L. Bonnemain, personal communication). For example, the spatial organization of the phloem in the culm nodes of grasses is far from being clearly understood.

Sieve tubes cannot be studied beyond a few elements by means of longitudinal sections. The degree of extension of individual files of sieve elements is consequently poorly understood, as are the junctions of these files between adjacent organs.

The technique of xylem casts has been applied to freeze-dried stem segments of squash (*Cucurbita maxima*), a species characterized by the large diameter of sieve pores (2 to 4 µm). The stems were previously subjected to an *in vivo* treatment designed to block the formation of callose and prevent the sealing of sieves. The sieve tubes of the external and internal phloem were cast at the same time as those of the xylem in stem segments around 20 mm long (Plates 5.7 and 1.9). Nevertheless, it is not certain that the elastomer followed the route of the sieves. A slight imprint of pores was visible in transmission light, but the lumina of their pore canals were not cast. It is thus possible that the elastomer has penetrated through tears in the wall, as is suggested for parenchyma cells.

PERSPECTIVES

The preceding examples mark the present limitations of the use of casting techniques such as those we have implemented. This technique is well adapted to the exploration of large intracellular and intercellular spaces, vessels, and canals. However, it has uncertain results when the casting agent penetrates intracellular spaces isolated by a primary wall that is supposed to be continuous, as are the tracheids and the parenchyma cells in which the cytoplasmic content has dried up, and as are the sieve elements of phloem in which the pores are sealed. Casting in these intracellular territories has to be mastered through a methodical study of the behaviour of the primary wall under the effect of desiccation and pressure.

In this regard, we should mention a recent improvement in a promising variant technique, intermediate between casting and impregnation, which allows examination under SEM of casts of cells fixed in the living state (Kitin et al., 2001). It is presently limited to tissue blocks of a few millimetres. Its adaptation to larger samples could make it a technique of choice for the study of phloem and fragile tissues such as meristems.

The reader may regret the absence of even a mention of applications to the examination of grafts and of wood altered by vascular fungi. Those are gaps that have not yet been filled, for lack of adequate plant material and time. But it is probable that the establishment of vascular junctions between stock and graft and the progression of parasite hyphae into the xylem are two processes the dynamics of which could usefully be examined by the technique of casting.

Shortly before the publication of this book, we did find an opportunity to observe by means of microcasting three ligneous species (citrus tree, maple, palm) attacked by three different pathogenic vascular fungi (*Phoma* sp., *Verticillium* sp., *Fusarium* sp., respectively).

Among many hundreds of vessels simultaneously cast in samples of metaxylem for the palm and wood for the trees, vessels with more or less damaged secondary walls and vessels with intact walls may easily be distinguished, some of them containing fungal hyphae and some not. Any wall damage necessarily modifies the surface state of the casts and results in lateral leaks of the elastomer. Under light transmission, the fungal hyphae embedded in the elastomer are clearly visible along vessel casts. Numerous tyloses forming spherical holes in cast were observed in the maple vessels only, which partly or almost completely obstructed their lumina.

These gradual symptoms of disease could allow us to follow accurately and easily the radial and longitudinal spread of the fungus in the vessels of a trunk or branch as early as necessary.

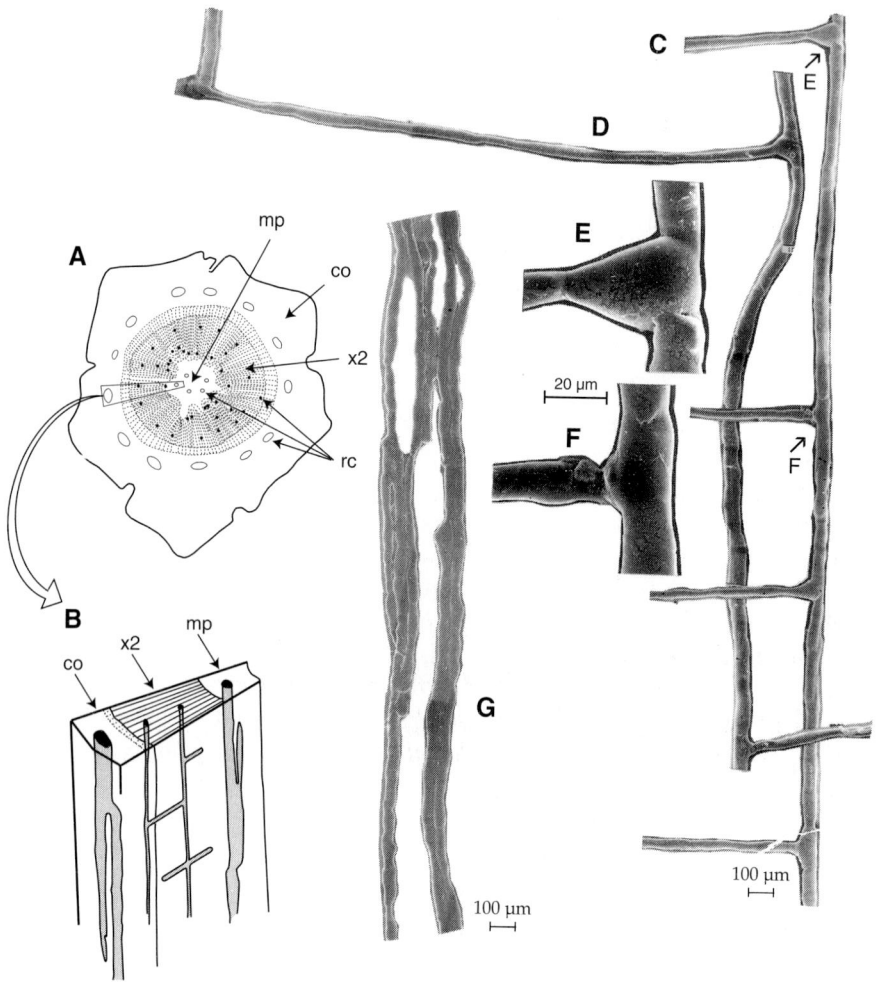

Plate 5.1. Resin canals in the wood and in the medullary parenchyma of *Pinus mugo*. A. Cross-section of a leafy branch. The resin canals (rc) form in the cortical parenchyma (co) and medullary parenchyma (mp) (diameter of canals, 70–100 μm) and in the axial and radial parenchyma of the wood (diameter, 30–50 μm). B. Drawing of the three types of canals. C, D. Cast segments of wood canals; the axial and radial branches are often anastomosed at right angles. E, F. Enlargements of two junctions. G. Segment of cast of medullary canals: the parallel branches are often anastomosed (SEM photos).

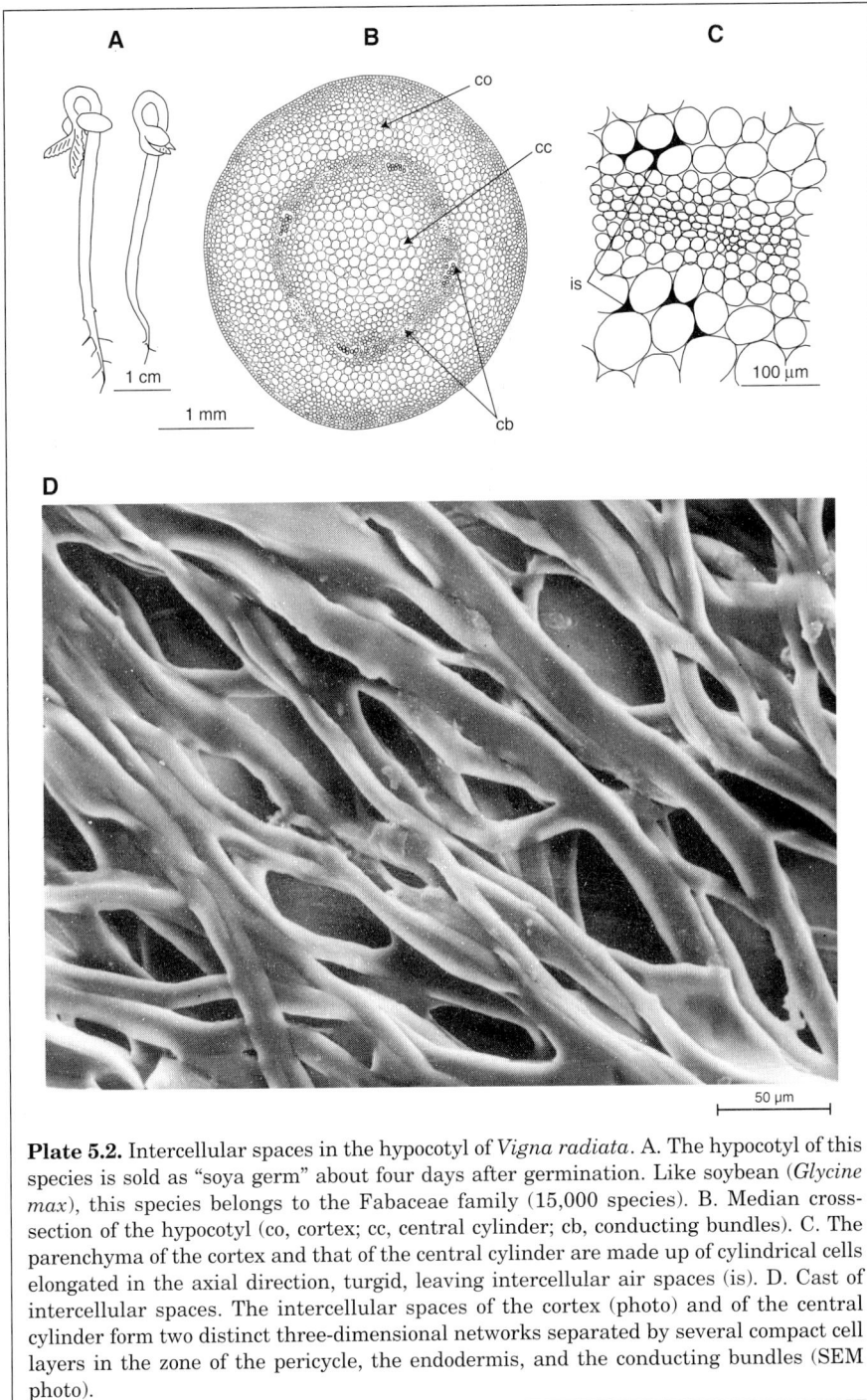

Plate 5.2. Intercellular spaces in the hypocotyl of *Vigna radiata*. A. The hypocotyl of this species is sold as "soya germ" about four days after germination. Like soybean (*Glycine max*), this species belongs to the Fabaceae family (15,000 species). B. Median cross-section of the hypocotyl (co, cortex; cc, central cylinder; cb, conducting bundles). C. The parenchyma of the cortex and that of the central cylinder are made up of cylindrical cells elongated in the axial direction, turgid, leaving intercellular air spaces (is). D. Cast of intercellular spaces. The intercellular spaces of the cortex (photo) and of the central cylinder form two distinct three-dimensional networks separated by several compact cell layers in the zone of the pericycle, the endodermis, and the conducting bundles (SEM photo).

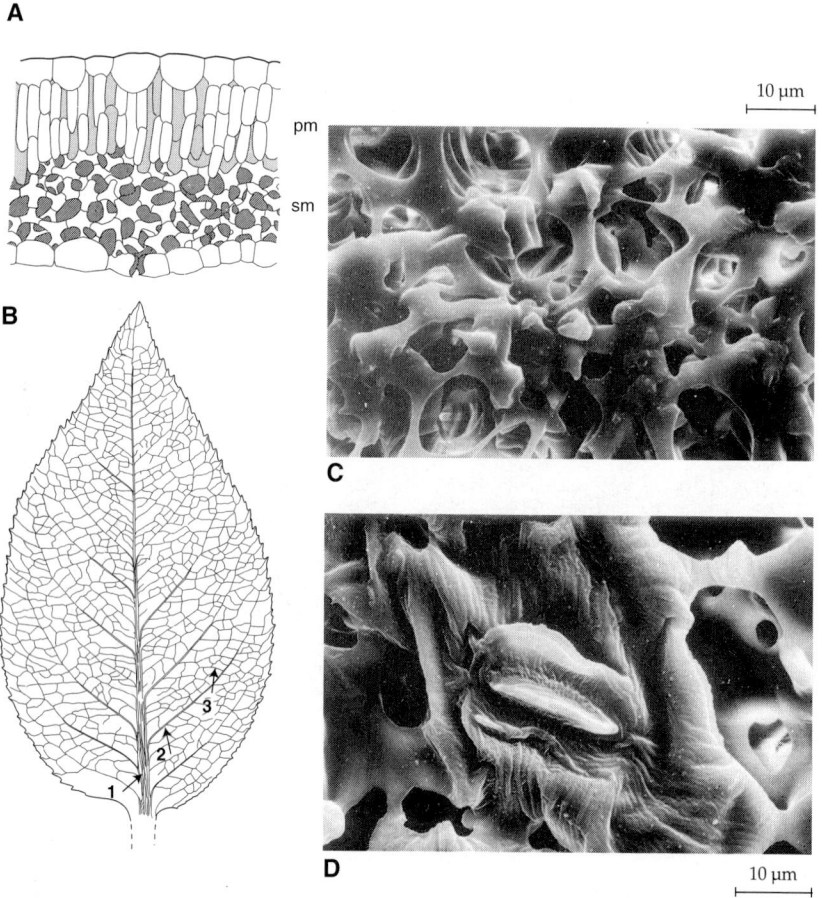

Plate 5.3. Empty space of mesophyll of a rose leaflet (fresh tissue). A. The air occupies a large space between the cells of the palisade mesophyll (pm) (upper side) and of those of the spongy mesophyll (sm) (lower side). B. The continuity of this empty space is interrupted in the limb by veins of order 1, 2 and 3. C. Cast of the empty space, seen from the upper side: the cavities are occupied by palisade mesophyll. D. The same, seen from the lower side. In the centre, there is an imprint of the inner surface of a stoma (SEM photos).

A

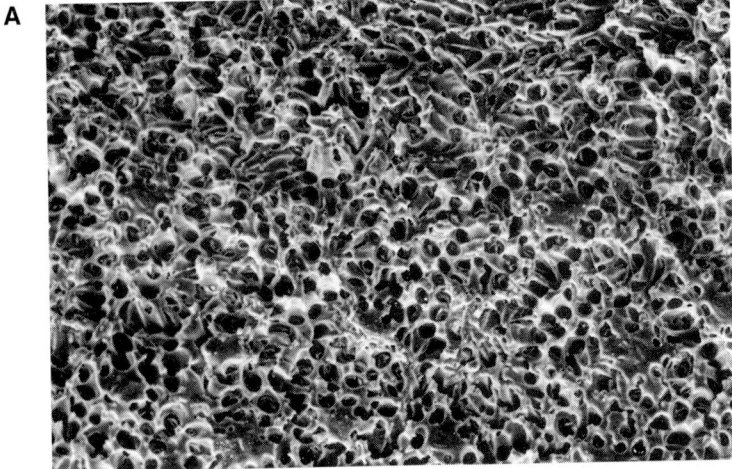

100 µm

B

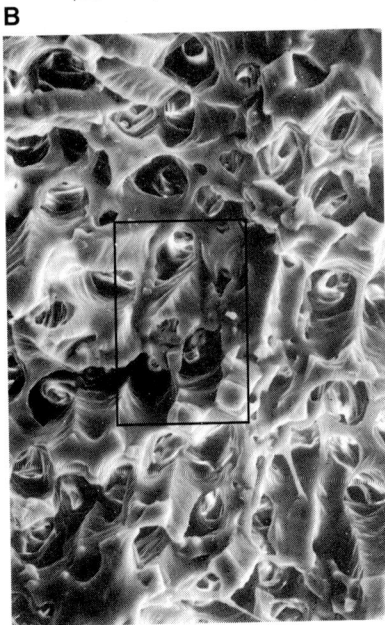

50 µm

C

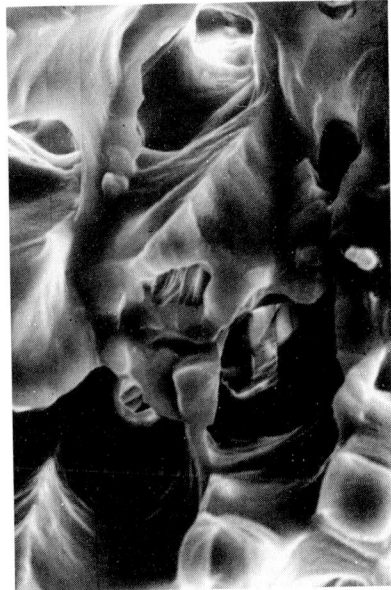

10 µm

Plate 5.4. Cast of the empty space of the mesophyll of a tomato leaflet (fresh tissue). It is thicker than that of the rose leaflet and interrupted by the proximal part of the veins of order 1 and 2, but continuous along the edge of the leaflet. A. Upper side. The cylindrical cavities are occupied by the palisade mesophyll. B, C. Enlargements of this side (SEM photos).

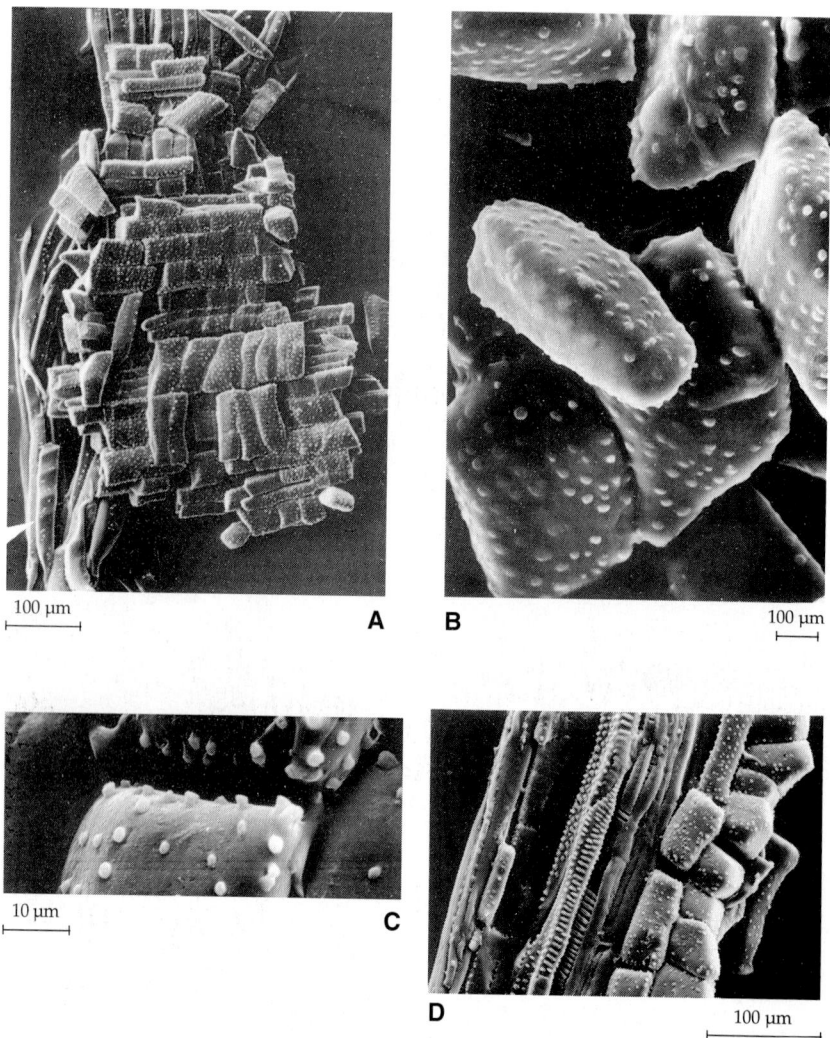

Plate 5.5. Radial parenchyma in secondary xylem of tomato plant. A. All the cell files of a ray adjacent to a vessel have been cast, at first through the vessel-ray pits then probably through ray-ray pits or tears in the wall. At lower left is an axial parenchyma cell (arrow). B. The cast cells of an aggregate are easily detached from one another. C. Ray-ray pits, few in number and small in diameter. D. Perivascular cast. Vessel elements at left. Axial parenchyma at centre. Ray parenchyma at right (SEM photos).

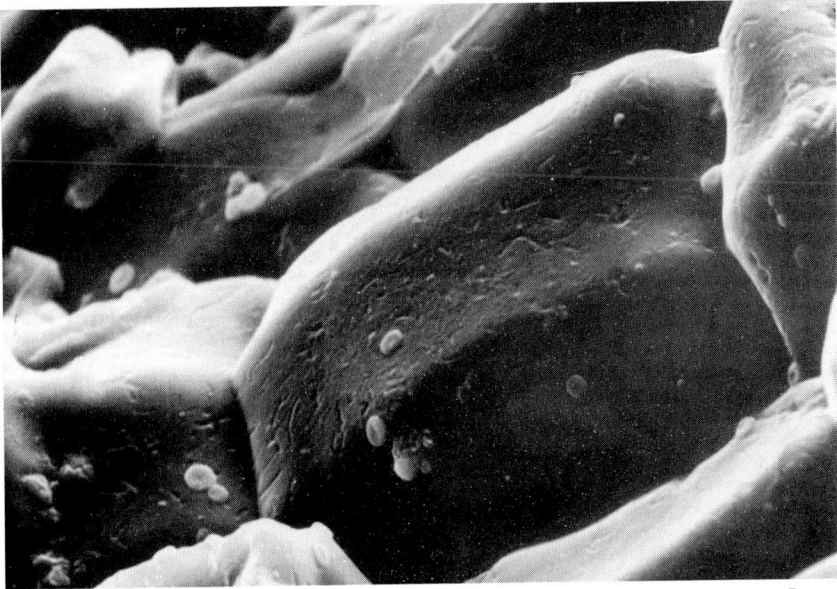

A   5 µm

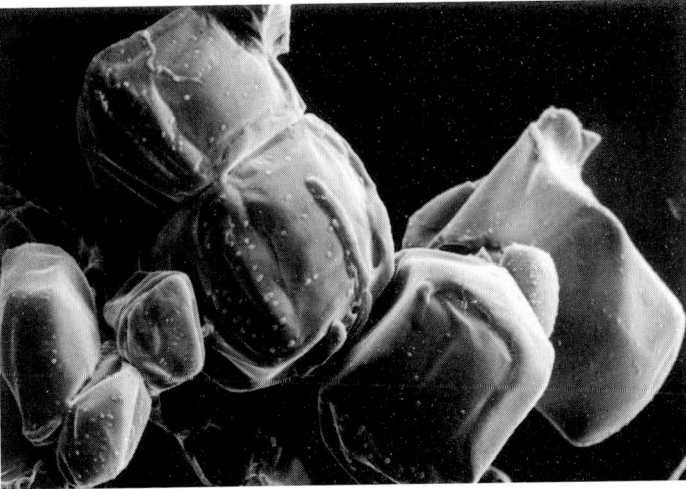

B

100 µm

Plate 5.6. A. Epidermal cell of tomato rootlet. Cast after freeze-drying. The grooves on the cast surface could be the imprints of desiccated cytoplasmic organelles stuck on the inner surface of the cell. The flattened circular forms are not identified. (Could they be bacterial contamination deposited on the cast?) B. Some cells cast in the medullary parenchyma of a rose stem. This parenchyma was dead before the cast was taken (SEM photos).

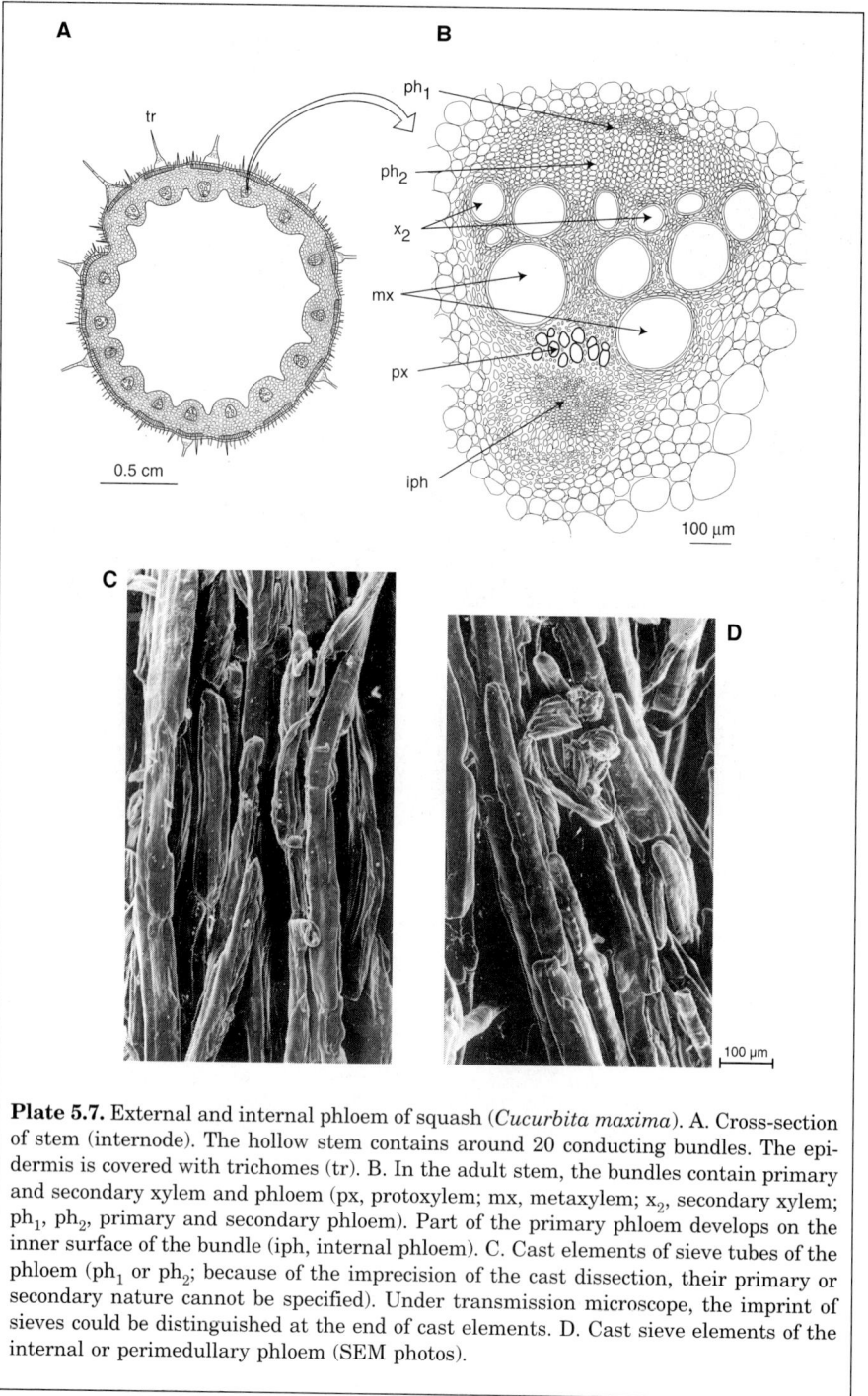

Plate 5.7. External and internal phloem of squash (*Cucurbita maxima*). A. Cross-section of stem (internode). The hollow stem contains around 20 conducting bundles. The epidermis is covered with trichomes (tr). B. In the adult stem, the bundles contain primary and secondary xylem and phloem (px, protoxylem; mx, metaxylem; x_2, secondary xylem; ph_1, ph_2, primary and secondary phloem). Part of the primary phloem develops on the inner surface of the bundle (iph, internal phloem). C. Cast elements of sieve tubes of the phloem (ph_1 or ph_2; because of the imprecision of the cast dissection, their primary or secondary nature cannot be specified). Under transmission microscope, the imprint of sieves could be distinguished at the end of cast elements. D. Cast sieve elements of the internal or perimedullary phloem (SEM photos).

Part II
The Technique of
Microcasting

Introduction

The application of casting to plant histology is an example of those numerous and profitable technological "odd-jobs", sometimes "breakthroughs", so common in research laboratories, where the idea is to find a principle, an accessory, or a product (in this case, a casting agent used for industrial purposes), and to apply it to serve new functions.

In fact, the principle of casting is simple and well known, and the idea of materializing the lumina of vessels so as to follow their course and ramifications in a tissue or organ more easily and to understand their organization in three dimensions was not novel. It was apparently first discovered and developed in the field of animal and human anatomopathology and it is still used for research and educational purposes (Wolfe, 1962; Murakami et al., 1973; Aharinejad and Bock, 1993; Okada and Ohta, 1994; Hossler, 1997). Remains are still found of early attempts to understand and explore living structures by means of casting, For example, the museum of a Neapolitan chapel displays in glass cases the skeletons of a man and a woman, enveloped in the net of their own complete circulatory systems, preserved by the injection of an unknown casting agent, probably during the 18th century!

The same principle was applied for approximately the past twenty years to the study of the simpler system of xylem vessels. The belated pioneers in this subject did not make the first casts but extracted them from sheets of plywood in which vessel casts were formed during gluing under pressure (Saiki et al., 1975; Okumura et al., 1976). When vulcanizing and cross-linking polymer compounds were commercially produced, it became possible to make and replicate numerous experimental casts in wood (Taneda et al., 1979; Smulski, 1980; Stieber, 1981; Gardner and Taylor, 1983; and more recently Fujii, 1993a and b; André, 1993, 1995, 1996a and b, 1998, 2000; Mauseth and Fujii, 1994; Fujii and Hatano, 1996; Prat et al., 1997; André et al., 1999; Fujii et al., 2001; Kitin et al., 2001).

Until about 1985, most of the possible applications of this technique of studying vessels were foreseen, but only as projects: e.g., the observation of the track and junctions of vessels, the measurement of their length, the examination of vessel elements *in situ* in their natural positions with respect to one another, the study of the permeability of the primary walls of pits. The studies of this period, limited to casts of linear vessels of secondary formation, soon came to an end.

During the 1980s, exploration of the vascular system also proceeded in directions other than casting. Two adaptations of the more classic technique of thin sections to the study of vessel tracks deserve to be mentioned here. The first consisted of using successive cross-sections in a small block of wood the size of a matchstick. On each section, the coordinates x, y of each vessel section were recorded in order to reconstruct the vessel forms from one section to the next according to axis z, by means of a three-dimensional drawing in perspective view (Zimmermann, 1982). The second used fine plates of wood 1 mm thick, sliced parallel to the wood grain and made translucent by a physicochemical process of impregnation. The vessels present in each plate were then coloured and observed under transmission light (Finck, 1992). These processes were laborious and their results have limited uses, but they indicate the efforts made to widen our understanding of conducting tissues.

After an interval of about ten years, the casting technique reappeared curiously at the same time in two distant parts of the world. In the 1990s, Fujii took up the researches of his predecessors, attempting mainly to identify woody species from morphological characters of their wood vessels revealed by casting. At the same time, and independently of all the studies on this subject, we proposed the use of vascular casts to locate hypothetical branched vessels in Dicotyledon wood, i.e., to locate triperforate elements. After our technique was progressively perfected, it proved to be highly useful for this type of research and sorting, leading for example to the discovery of heterogeneous vessels in Dicotyledons, then in Monocotyledons. When this step was taken, the major utility of casting appeared to be of greater interest with the discovery of complex vascular structures, such as the nodal ramifications of metaxylem in Monocotyledons.

This new technique has its most useful applications in the field of morphology and mostly of spatial organization of certain tissues. Certainly the cast images given by SEM are often striking, highly informative, and pedagogical. It complements the technique of serial sections, which remains irreplaceable.

Another aspect that should be mentioned is the progress of confocal microscopy developed in plant anatomy in the 1990s (Knebel and Schnepf, 1991; Gray et al., 1999). Its principle consists of scanning, with the point image of a laser source, successive parallel planes in the total thickness of a tissue made fluorescent by impregnation. The light reflected is recorded and treated to provide a synthetic image of the volume explored: at present,

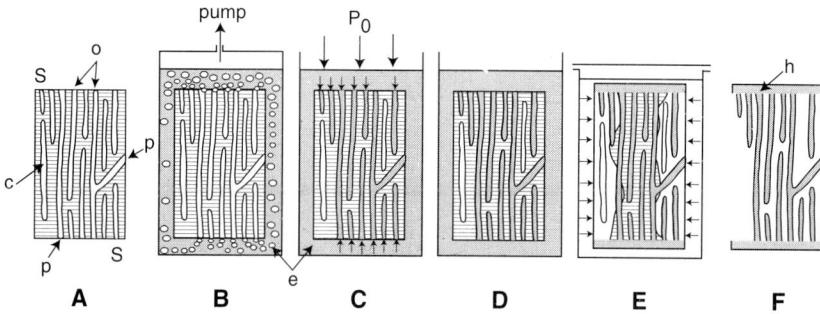

Fig. II.1. The steps of microcasting. A. The segment of organ or tissue to be cast is illustrated as a porous block. Some air spaces are indicated, some closed (c), others open (o) by one or two pores (p) on the sections (S). B. Vacuum degassing after immersion in the liquid elastomer (e). The air in the open spaces escapes. C. The re-established atmospheric pressure (P_0) forces the liquid to penetrate the open spaces: this is injection. The closed spaces are not cast. D. When the injection is completed, the elastomer is hardened by cross-linking. E. The flanks of the segment are separated from the outer layer of the elastomer. The tissues are destroyed by acid and oxidizing solutions. F. The casts of open spaces held in place by heels (h) are released.

the volumes explored are of cellular dimensions and the resolution of details of the order of one micrometre.

These new tools help us to link the structures and functions of tissues more closely in our understanding.

The principal steps of the microcasting technique are described in detail in the following sections and summarized in Fig. II.1:

—preparation of plant material,

—injection of casting agent,

—properties and crosslinking of the casting agent,

—chemical destruction of plant tissues,

—manipulation and examination of casts.

Preparation
of Plant Material

The choice of species, organs, and tissues, like the choice of the size of the sample to be prepared, depends on the research aims and the nature of tissue spaces to be studied. As shown in Part I, this technique is applicable to a wide range of plant samples, from the hardest, such as wood, to the most fragile, such as the primary and herbaceous tissues. The field will undoubtedly extend much further with new improvements in technology.

The preparation of samples, which largely determines the quality of the final casts, consists of locating anatomically the spaces to be cast, intra- or intercellular, and rendering them accessible to the casting agent. Since the requisite destruction of the plant tissue at the end of the operation leads to the loss of anatomical markers that are essential for the interpretation of casts, a precise and complete descriptive profile must first be established for each sample.

DESCRIPTIVE PROFILE

The following list of useful details to be recorded is given as an indication, in light of our own experience:
- Name of species and variety.
- Place and date of collection. Mode of preservation after collection. Drying.
- Precise drawing, at a useful scale, of organs from which the segments to be cast are sliced. Indication of insertions of adjacent organs, scars of organs detached previously. Phyllotaxy of the species. Indication of the limits of growth units.
- Possible photographic complement. Photocopy of flat organs.
- Precise drawing or photograph of cross-sections. Location of primary and secondary tissues. Estimation of diameters of air spaces to be cast.

PLANES OF SECTION

The sawn cross-sections of lignified organs are carefully trimmed with a razor in order to open the surface pores, which are the routes through which the casting agent enters. The lower and upper ends of the cylindrical segments (internodes) are sawn along different distinctive angles.

Longitudinal separation by splitting along the wood grain is generally preferable to sawing.

The size of the samples to be cast is limited by that of the injection chambers (see p. 120). In view of the available material, we have been able to make vascular casts of 50 to 100 cm length on clematis stem and of 10 cm diameter on a segment of palm stipe as examples of extreme size. The problems posed by the manipulation and preservation of large casts are addressed on pages 132. Generally, our samples were of medium size, measuring 3 to 8 cm in their largest dimension.

Short entire vessels may be found isolated between the two sections of a stem or branch segment. They are inaccessible to the casting agent unless the segment was opened up by a third median section. In this case, the successful casting of whole vessels requires that the two parts of the segment be strongly and exactly maintained in their relative positions during the cross-linking step. Different processes could be proposed, which we have not yet implemented.

TREATMENT OF SPACES TO BE CAST

Once the pathways of entry have been cleared for the casting agent, the plant material must be prepared so that the casting agent (1) occupies all the spaces to be cast and (2) hardens by cross-linking under optimal conditions. The first condition demands that all the liquid content be taken out from the spaces to be cast without deforming the tissues, i.e., removing the water from the sap-conducting vascular system and the resins, gums, and latex from the various ducts conducting these substances. The second condition aims to eliminate any substance that could contaminate and affect the cross-linking catalyst.

- The *water* is evacuated by partial or total desiccation of the organ, before or after the sectioning of segments. Air-drying at constant weight at ambient temperature is the simplest procedure when the tissues are lignified and do not deform easily. Non-lignified tissues, which have a higher water content, and primary tissues in general, are freeze-dried or dried by the critical point drying technique.[1] In parenchyma,

---

[1]In the freeze-drying technique, the tissue water is frozen (at $-40°$ to $-180°$, depending on the equipment.) The ice formed in the cells is then sublimated under vacuum. In the critical point drying technique, the tissue water is displaced by acetone, then the acetone is displaced by liquid CO_2 at a given pressure and temperature. Then, the liquid CO_2 is slowly vapourized at the temperature and pressure of its critical point.

there are intercellular spaces (in the mesophyll of leaves, petals, certain stems, etc.) in which casting is possible on fresh tissue without prior treatment. The presence of cytoplasmic water is not an obstacle to the casting itself.

- The *resin* is eliminated after desiccation by repeated washes in turpentine and mixtures of acetone and ethyl alcohol, followed by air-drying. The case of *gums* and *latex* is not addressed.

- Inhibition of the cross-linking reaction has been observed in experiments when the casting agent is injected into the primary vessels of growing shoots (especially Poaceae). An ethanol extraction at 50°C, followed by freeze-drying, largely eliminates this defect, but the nature of the substances affecting the catalyst was not determined (see p. 125, properties of the elastomer).

OBSTACLES TO THE PROGRESSION OF THE CASTING AGENT THROUGH THE TISSUES

Generally, the primary walls that are continuous around each cell are permeable to water but impermeable to the casting agent (liquid silicone). Nevertheless, depending on the nature of the cells, their physiological function, and their age, the primary walls have various degrees of porosity, following natural alterations of their chemical structure. In addition to this degrading chemical process, physical effects of drying of tissues before the casting and pressure variations during the operations could lead to tearing of the primary wall.

The progression of the casting agent through the tissues is totally blocked by intact primary walls only. For example, when it progresses into lumina of two adjacent vessels of young wood, the liquid invades the chambers of their respective pits but does not cross the double primary walls that separate them. All other things being equal, this barrier is more easily broken in dead wood. The double primary wall is often porous or perforated between the adjacent tracheids of Gymnosperms and Pteridophytes. Finally, and this is common but difficult to predict, slight tears in the primary wall could be responsible for the entry of elastomer into cell lumina of parenchyma adjacent to vessels, for example, as well as for leaks of small amounts of elastomer, as revealed in the examination of numerous plates in Part I of this book. This does not affect the quality of the casts.

The lumina of aged vessels of Angiosperms are often partly or totally sealed by tyloses. These are outgrowths of living cells, adjacent to vessels, which enter into vessels through pits and enlarge further there. Even after drying, their primary wall is an obstacle to the progression of the casting agent.

Injection
of the Casting Agent

PRINCIPLE OF INJECTION

As indicated in Fig. II.1B and C, the injection occurs *simultaneously* in all the empty spaces, except the closed ones, of the samples immersed in the casting agent. To do this, the air present in the empty spaces of tissues is first evacuated by a vacuum pump (degassing step). As soon as the atmospheric pressure is re-established on the surface of the liquid, it forces the elastomer to penetrate the pores of the samples (this is strictly speaking the injection step).

The properties of the casting agent are described in the next section of Part II.

DEGASSING APPARATUS

The equipment is made up of a vacuum chamber with a mercury manometer that is connected either to a vacuum pump or to the atmosphere. The vacuum chamber commonly used is tubular, but any shape may be used according to sample sizes. The plant segments are placed in a tubular polythene sheath heat-sealed at the bottom and immersed in a sufficient quantity of casting liquid. The sheath is kept suspended in the tube (Fig. II.2A).

DEGASSING

The tube and its contents are progressively put under vacuum (Fig. II.2B), with constant control of foam production. As the plugging of the chamber is not perfectly airtight and as a slight flow of water vapour may escape from moist or partly dried samples, the residual pressure P_1 does not fall below 10^{-3} and 10^{-4} P_0, P_0 being the atmospheric pressure (Fig. II.3). This operation need not be prolonged beyond 20 minutes.

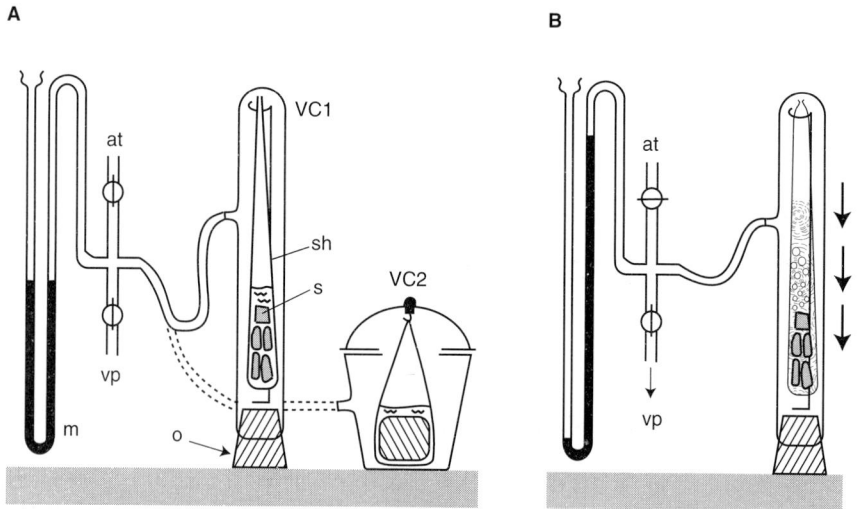

Fig. II.2. Degassing apparatus. A. VC, vacuum chambers. VC1 is a glass tube (length 100 cm, internal diameter 5 cm). VC2 is a wider vacuum chamber. sh, plastic sheath; s, immersed samples; vp, vacuum pump; at, atmosphere; m, manometer; o, obturation. B. Degassing. The foam that forms can be broken up by a few vigorous taps of the tube (arrows).

INJECTION

The atmospheric pressure P_0 is gradually re-established in the tube. It may be assumed that the penetration of the liquid, which immediately begins in the various ducts, follows the Hagen-Poiseuille law (Fig. II.3).

The sheath containing the segments is taken out of the tube and placed in a deep freezer during the estimated duration of the injection, in order to block the process of cross-linking of the casting agent, i.e., the process of hardening (see next section). During the injection, the samples must be kept immersed.

DURATION OF INJECTION

In the absence of empirical measures of the speed of penetration of the casting agent in calibrated vessels, the durations of injection have been calculated on the basis of the Hagen-Poiseuille law applied to the following two cases: the first is that of a capillary of a given length and diameter open at one end; the second is that of a hollow volume of parallelepipedic shape pierced with a small opening.

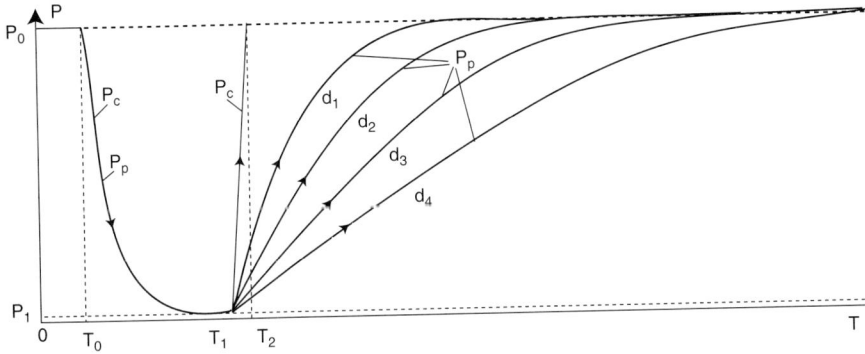

Fig. II.3. Hypothetical variations of pressure inside the chamber (P_c) and in the open porosity (P_p) of samples during degassing and injection. Degassing between T_0 and T_1: drop in pressure from P_0 to P_1. Re-establishment of pressure P_0 in the chamber between T_1 and T_2. Rise of pressure during the injection in the lumina of 4 vessels of decreasing diameter ($d_1 > d_2 > d_3 > d_4$).

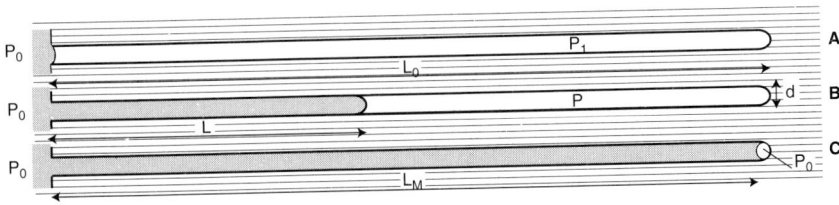

Fig. II.4. A capillary before, during, and after injection (see text).

• Filling of a capillary

This fictitious vessel, of length L_0 and diameter d, initially contains air at very low pressure P_1. It is opened into a mass of liquid at atmospheric pressure P_0 (Fig. II.4A). When the liquid column of viscosity η reaches the length L, the internal air pressure reaches the value P (Fig. II.4B). At the asymptotic end of the filling, L_M will be the maximal length of the liquid column and P_0 the final internal pressure (Fig. II.4C). We suppose that in the liquid column a linear pressure gradient is constantly established between its two ends.

Under these conditions, the liquid flow $d\vartheta/dt$ is proportional to the pressure gradient in the column equal to $(P_0 - P)/L$

$$\frac{d\vartheta}{dt} = \frac{\pi d^2 dL}{4dt} = \frac{\pi d^4}{128\eta} \frac{(P_0 - P)}{L} \qquad \text{(equation 1)}$$

It is supposed, moreover, that the air behaves like a perfect gas.

$$P_1 L_0 = P (L_0 - L) = P_0 (L_0 - L_M) \qquad \text{(equation 2)}$$

Thus
$$L \left(\frac{L_0 - L}{L_M - L} \right) \cdot dL = \frac{P_0 \cdot d^2}{32\eta} \cdot dt$$

and, after integration, the duration of filling as a function of the length of the liquid column is expressed as follows:

$$t = \frac{32\eta}{P_0 \cdot d^2} \cdot \left[\frac{L^2}{2} - (L_0 - L_M) \cdot L - L_M (L_0 - L_M) \cdot \ln\left(\frac{L_M - L}{L_M} \right) \right]$$

with
$$L_M = L_0 \cdot \left(\frac{P_0 - P_1}{P_0} \right)$$

The graph in Fig. II.5 is based on the following numerical values: $P_0 = 10^5$ Pa; $P_1 = 10^2$ Pa; $L_0 = 10^{-1}$ m; $\eta = 4$ Pa $\cdot$ s; $d = 100, 50, 25,$ and $12.5 \cdot 10^{-6}$ m.

According to these hypotheses, vessels of diameter greater than 25 µm are full at the end of 3 h, and those of diameter 12.5 µm are only half full.

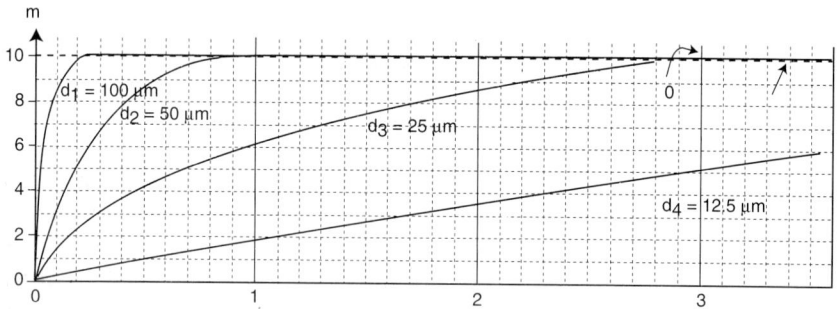

Fig. II.5. Progression of the elastomer in capillary tubes of decreasing diameters ($d_1 = 100$ µm, $d_2 = 50$ µm, $d_3 = 25$ µm, $d_4 = 12.5$ µm). L is the length reached in time t, h = hours.

• Filling of a pierced volume

This volume represents a fictitious empty cell adjacent to a vessel, such as a cell of radial parenchyma, and communicating with the vessel by the orifice of a single perforated pit. The initial conditions are comparable to those described above. The volume V_0 filled with air at pressure P_1 is exposed to the liquid itself at pressure P_0. The liquid penetrates through a duct of diameter d and length Δ. It is supposed that the pressure gradient between the ends of the duct is linear (Fig. II.6).

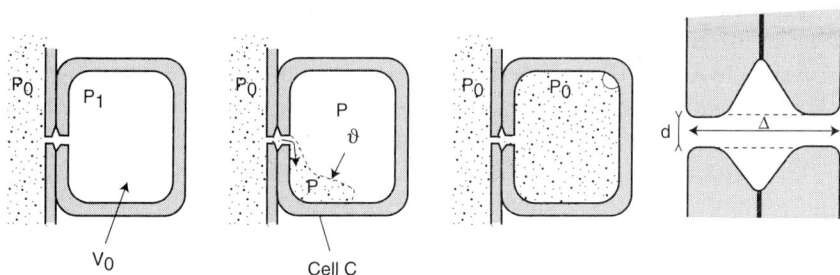

Fig. II.6. A pierced volume before, during, and after injection (see text). Right: perforated pit.

Equations 1 and 2, analogous to the preceding ones, are as follows:

$$\frac{d\vartheta}{dt} = V_0 \frac{d\omega}{dt} = \frac{\pi d^4}{128\eta} \cdot \frac{P_0 - P}{\Delta} \qquad \text{(equation 1)}$$

$$P_1 \cdot V_0 = P \, (V_0 - \vartheta) \qquad \text{(equation 2)}$$

where $\omega = \dfrac{V_0 - \vartheta}{V_0}$.

Thus: $\omega + \dfrac{128\eta \cdot \Delta \cdot V_0 \cdot \omega}{\pi d^4 P_0} \cdot \dfrac{d\omega}{dt} = \dfrac{P_1}{P_0}$

and, after integration:

$$t = \frac{128\eta \cdot \Delta \cdot V_0}{\pi d^4 P_0} \cdot \left[\frac{\vartheta}{V_0} - \frac{P_1}{P_0} \cdot \ln\left(\frac{1 - \vartheta / V_0 - P_1 / P_0}{1 - P_1 / P_0} \right) \right]$$

The duration of filling is calculated for a given filling rate ϑ / V_0.

The numerical values are the following:

$P_0 = 10^5$ Pa; $P_1 = 10^2$ Pa; $V_0 = 50 \times 40 \times 50$ μm^3 or 10^{-13} m^3; $d = 2 \cdot 10^{-6}$ m;

$\Delta = 5 \cdot 10^{-6}$ m; $\eta = 4$ Pa $\cdot$ s.

The time required to reach a filling rate of 99% is about 50 seconds.

• **Remarks**

The durations thus calculated are based on simple hypotheses, especially with respect to the linear pressure gradient, and do not take into account the real forms and wrinkles of volumes and surfaces. They are used for base estimations of real durations. In practice, the mean time of injection into vessels is about 10 to 15 h at a temperature of –18°C.

According to the calculations, the cell fills three times as quickly as a segment of the same volume of the finest capillary (d = 12.5 μm): under these conditions, the casting of aggregates of several tens of radial parenchyma cells (see Plate 5.5) seems less surprising.

INJECTION OF COLOURED ELASTOMER

The elastomer can be coloured by insoluble pigments made of fine particles of diameter less than one micrometre (supplied by Dow Corning S.A.). Pigments can be used for example to mark the vessel casts of two organs adjacent to one another. To do this, the elastomer is injected in one of the organs after the cut ends of the other organ are masked (Fig. II.7).

The pigments are visible in the transparent elastomer under transmission light microscope. Thus, their accumulation allows us to locate small perforations traversed by the elastomer but not by the particles (Fig. II.8).

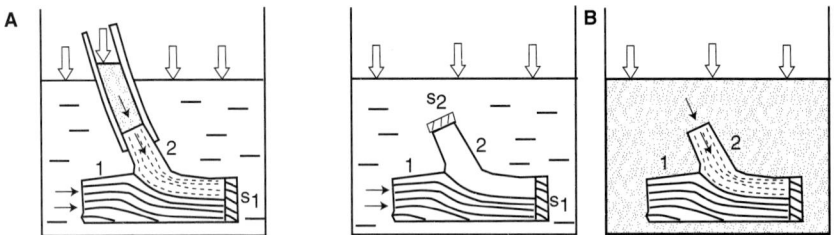

Fig. II.7. Injection of pure elastomer in the segment of axis 1 and of coloured elastomer in the segment of axis 2 in the chamber after degassing. A. Masking of s_1 and simultaneous injections. B. Masking of s_1 and s_2 and injection in axis 1. Crosslinking, then injection in axis 2. The masks are composed of silicone cement.

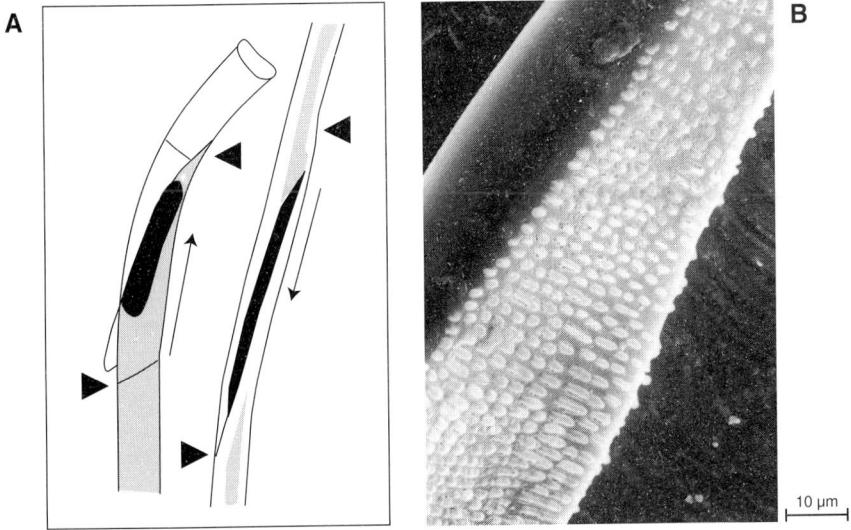

Fig. II.8. A. Injection of the elastomer charged with pigment in the metaxylem of a bamboo (*Phyllostachys aurea*), Injection in the direction of the arrows. The pigment particles are retained along the oblique planes of the perforations. B. Cast of the surface of a plane of perforations traversed by the elastomer but not by the particles.

Properties and Crosslinking of the Casting Agent

Needless to say, no commercially available casting agent is specifically designed for histological purposes. The authors cited in the introduction have used and continue to use a limited number of products. Some are liquid monomers to which has been added a polymerization catalyst (monostyrene→polystyrene; mixture of a diacid and a dialcohol→polyester), and others are liquid polymers that can be crosslinked at low temperature (silicone elastomer, which we use). Two authors have made the surprising choice of molten polyethylene at temperature between 170° and 220°C. The physical properties of casting agents (viscosity, temperature of use in liquid phase, resistance to shocks in the solid state) and their chemical properties (chemical inertia with respect to strong acids and oxidizing solution) determine the field of application of the technique and the quality of the casts.

PROPERTIES OF SILICONE ELASTOMER RTV 141

The room temperature vulcanizable (RTV) range of Rhone-Poulenc, France, is composed of liquid polysiloxanes produced for various industrial uses (e.g., protection of surfaces, water repellence, electric insulation, glues, casts, moulding, protection against fire).

The product RTV 141 consists of two components, A and B. The first is a polysiloxane with long chains, bearing vinyl and methyl radicals, supplemented by a catalyst (chloroplatinate). The second is a polysiloxane with short chains and H radicals. They are thoroughly mixed with a spatula before use in a proportion respectively of 10:1. The two liquid compounds are colourless, odourless, non-toxic, and insoluble in water. The mixture of the two has a density similar to that of water (d = 1.02) and an initial viscosity of 4 Pa·s.

The hardening results from a reaction of addition without by-product (see Table 3). The speed of the reaction increases with the temperature (see Table 3). The hardened elastomer is transparent with a refractive

Table 3: Crosslinking reaction in RTV mixture

Structure of a polysiloxane	Reaction of addition			

	−20°C	+20°C	+50°C	+100°C
Temperature of crosslinking				
Duration of crosslinking	50d	2d	4h	2h
Linear shrinkage		< 0.4 %	< 1%	≤ 1%

Manufacturer's data (Rhone-Poulenc, 1988).

index of 1.40. The elastic limit of elastomer filaments constituted by the vascular casts seems to vary inversely with their diameter, as if the passage through the network of the finest pores has the effect of increasing the number of covalent bonds between the chains and, thus, the elasticity of the hardened elastomer.

The elastomer is chemically degraded by pure sulphuric acid, but it is inert in its hydrate H_2SO_4, $2H_2O$ (acid at 72% by weight). It is also degraded by gaseous and dissolved chlorine but inert in concentrated sodium hypochlorite solutions (see p. 131).

The catalyst may be contaminated by sulphurous compounds (vulcanized rubber, for example) and nitrogenous compounds, the chemical nature of which has not been specified by the manufacturer. Such compounds (amino acids?) raise some problems in the case of young, growing tissues. Their inhibitory effect on crosslinking catalysis, probably superficial, is increasingly perceptible as the casts become finer (see p. 117).

PRACTICAL CONDITIONS OF CROSSLINKING

When the injection is completed, the tissue segments immersed in the elastomer are brought back to room temperature, deposited on stiff paper, separated from each other, and heated in a drying oven at 50°C if the tissues are lignified. If they are fresh and slightly lignified, the samples are first maintained at 30°C until a gradual hardening occurs without tissue deformation and, after that, at 50°C to obtain a sufficient hardness, which can be felt to the touch.

The segments are then carefully detached from the support, partly freed of the outer elastomer layer, and rehydrated in warm water for a few minutes so that they can be pulled out completely without breaking the fragile parts.

It is useful to leave in place the outer layer of the elastomer that covers the cross-sections of segments. This layer will hold the ends of the filaments in their relative positions when the tissues are destroyed.

Chemical Destruction of Plant Tissues

PRINCIPLE

The plant tissues, mainly the cell walls, are destroyed in two stages at room temperature and in a liquid medium. The first step is the hydrolysis of polysaccharides (cellulose, hemicelluloses) by sulphuric acid and the second is the oxidation of lignin by sodium hypochlorite. At the end of these two treatments, only the elastomer remains.

We will mention here the technique of Gardner and Taylor (1983), which consisted of dissolving only the middle lamella to preserve the walls of the vessel elements. According to these authors, the casting of vascular lumina was useful only to keep these walls in place.

HYDROLYSIS OF POLYSACCHARIDES

Each rehydrated and wiped cast sample is plunged into sulphuric acid H_2SO_4, $2H_2O$* and kept immersed under a small block of the same elastomer. The hydrolysis, very rapid in slightly lignified tissues, progresses from 1 to 2 mm per day in wood and ends with the softening and darkening of the entire bulk of segments. The acid must be changed in the case of large samples. The segments, which become fragile, are carefully transferred into a large volume of water until their density, which decreases with the dilution of the acid, allows them to float on the surface. The residual acid in the tissues is neutralized by a sodium bicarbonate solution until the total cessation of effervescence.

---

* The highly exothermic preparation of the hydrate H_2SO_4, $2H_2O$ is safely done by freezing the pure acid and the water in a deep freezer before mixing them. The acid is poured slowly on the ice, homogenizing it, the whole being frozen in an ice bath. For 1 litre of the mixture, 1180 g of the acid and 430 g of water should be measured out.

OXIDATION OF LIGNIN

The neutralized samples are gently and separately arranged in Petri dishes containing water. The principal morphological characteristics still preserved in the lignin are recorded according to the descriptive profile.

The oxidizing solution of sodium hypochlorite 2M (concentrated bleach at 48 French chlorometric degrees) is added drop by drop on the samples. It is useful to follow the gradual vanishing of the lignin and the appearance of the casts under a binocular microscope, e.g., for the purpose of taking photographs. The elimination of lignin leads to the subsidence of casts in which the filaments remain fixed at their ends by "heels" of elastomer. The casts are then carefully washed with a gentle jet of water.

A final test of coloration with a safranine solution, which is not fixed by the elastomer, can be used to detect the presence of residues of cell walls and to eliminate them by prolonging the hypochlorite treatment.

Figure II.9 summarizes the operations.

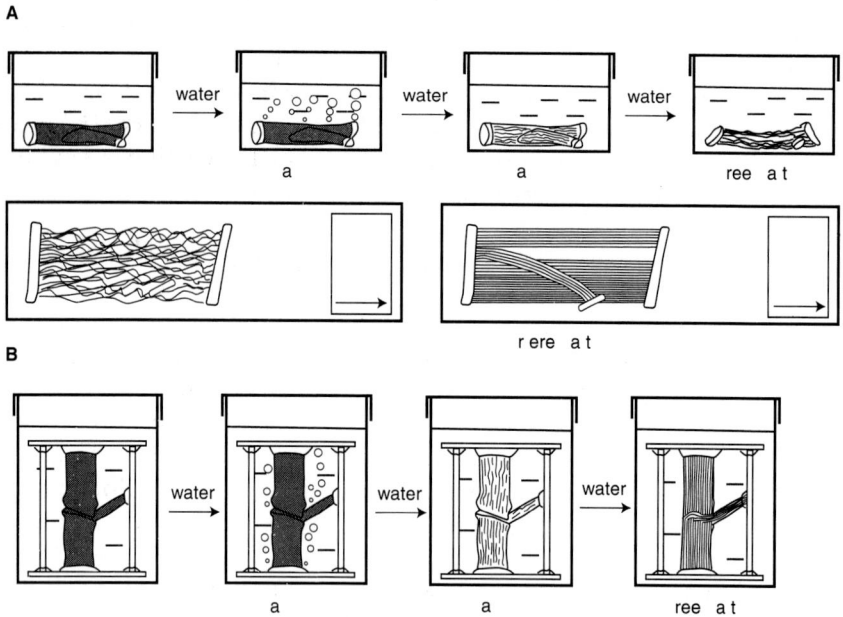

Fig. II.9. Destruction of plant tissues and separation of casts. A. The clean casts of small samples are spread out on a glass slide and the filaments are ranged in an anatomical sequence. The arrows indicate the basal-distal direction. B. Large samples can be fixed and stuck to a framework formed of glass plates and rods assembled with silicone cement before the destruction of tissues. The entire cast set-up is kept immersed in water (see Plate 2.11).

Manipulation of Casts

CHARACTERISTICS OF SILICONE ELASTOMER CASTS

Most of the casts are of vessels and intercellular spaces, i.e., filaments with a diameter between 5 and 400 μm (see for example Plates 1.9 and 5.2). The mechanical properties of the elastomer make it the only casting agent that makes it possible to obtain a very long and rather strong filament (possibly up to 1 m), provided the degassing apparatus is long enough. The filaments are elastic, flexible, and hydrophobic, then strongly subject to surface tension of pure water.

The sharpness of the details reproduced by casting is of the order of 0.1 to 0.2 μm, according to observations made on the cast of a finely photo-engraved electronic component used as a standard. Such a resolution is achieved because of the fluidity and the low surface tension (not specified by the manufacturer) of the liquid elastomer.

MANIPULATION

The different parts of each clean cast are arranged on transparent supports following the anatomical characters of the tissue samples, in order to facilitate later, extensive examinations. Glass slides or larger Perspex plates are used depending on the cast sizes.

The casts are always gently manipulated under a liquid film of a diluted solution of a wetting agent, using tools of microsurgery, microscissors, microscalpels, very fine tweezers, mounting needles, and magnifying binocular glass (Fig. II.10).

When the liquid film dries, the casts stick to their support and to one another. They can be preserved in this state for years. Some drops of the wetting solution must be added whenever they need to be manipulated again.

For optimal conditions of observation under magnifying glass and light microscope, transparent wetted supports are placed over a mirror and illuminated by means of optical fibres at a 45° angle (Fig. II.10). Particular details appear in low-angled light. It is difficult to describe the richness of

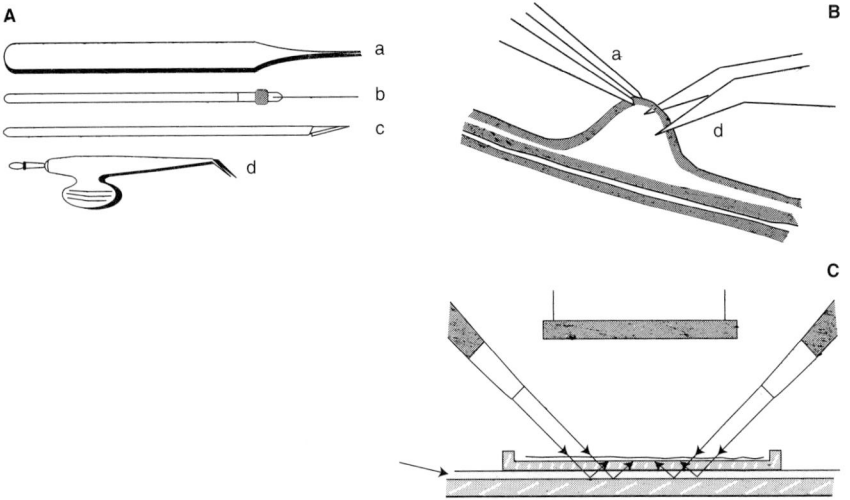

Fig. II.10. A. Tools used: (a) fine tweezers, (b) mounting needle, (c) microscalpel, (d) microscissors. B. Sectioning of a vascular cast. C. The object to be manipulated is spread out on a transparent support having a rim, which in turn is placed over a mirror (m). The light from optical fibres must be both direct and reflected. The object is visible above the support.

anatomical data revealed during the manipulation when the entire preparation is observed. A photograph can render only a very limited image.

PHOTOGRAPHY OF CASTS

The parts meant for *scanning electron microscope* are transferred on to stainless steel supports and spread out in a liquid drop of ethanolic or aqueous wetting solution. After they are arranged, the wetting solution is eliminated by aspiration. Gold-coating is generally done under 1.2 kV and 10 mA for 10 minutes and the observation under an acceleration voltage of 15 kV.

Examination under *light* microscope reveals the details embedded in the thickness of the transparent cast: the contour of simple perforations, the bars of scalariform perforations, the torus of gymnosperm pit pairs, the residues of protoxylem walls, possible coloured pigments, the hollow imprint of tyloses, and many other details.

References

Aharinejad S., Bock P. (1993). Casting with mercox-methacrylic acid mixtures causes plastic sheets on elastic arteries. *Scanning microsc.* 7 (2): 629–635.

Aloni R., Wolff A. (1984). Suppressed buds embedded in the bark across the bole and the occurrence of their circular vessels in *Ficus religiosa. Am. J. Bot.* 71: 1060–1066.

Aloni R., Pradel K.S., Ullrich C.I. (1995). The three dimensional structure of vascular tissues in *Agrobacterium tumefaciens*—induced crown galls and in the host stems of *Ricinus Communis L. Planta* 196: 597–605.

André J.P. (1993). Micromoulage des espaces vides intra-et intercellulaires dans les tissus végétaux. *C.R. Acad. Sci.* Paris 316: 1336–1341.

André J.P. (1995). Long vessel microcasting: Vessel ends and irregularly perforated vessel elements (poster) IUFRO 20th world congress. Tampere. Finland. *IAWA J.* 16: 5–6.

André J.P. (1996a). Investigations on circular vessels, by application of the vascular microcasting method (poster). Wood anatomy symposium. London. *IAWA J.* 17: 232.

André J.P. (1996b). Investigations on the vascular organization of the bamboos by application of the microcasting method (poster). Wood anatomy symposium. London. *IAWA J.* 17: 233.

André J.P. (1998). A study of the vascular organization of bamboos (Poaceae Bambuseae) using a microcasting method. *IAWA J.* 19: 265–278.

André J.P. (2000). Heterogeneous, branched, zigzag and circular vessels: unexpected but frequent forms of tracheary element files: description—localization—formation. In: *Cell & Molecular Biology of wood formation*. Savidge R., Barnett J., Napier R. (Ed.), Bios Scient. Publ. Ltd. Oxford chap. 31: 387–395.

André J.P., R. Prat R., Mutafschiev S., Catesson A.M. (1995). *3 D study of intercellular gas spaces in Mung bean hypocotyls.* (poster). 7th Cell Wall Meeting, Santiago de Compostela, Espagne.

André J.P., Catesson A.M., Liberman M. (1999). Characters and origin of vessels with heterogenous structure in leaf and flower abscission zones. *Can. J. Bot.* 77: 253–261.

Bierhorst D.W., Zamora P.M., 1965. Primary xylem elements and element association of Angiosperms. *Am. J. Bot.* 52: 657–710.

Böhlmann D. (1970) III. Die Abzweigungsverhältnisse bei *Quercus robur* und *Populus sektion Aigeiros. Allg. Forst-Jagd Zeitung* 141: 224–230.

Bugnon P. (1924). Contribution à la connaissance de l'appareil conducteur chez les Graminées. *Bull. Soc. Linn. de Normandie.* Mém. 26: 21–40.

Busby C.H., O'Brien T.P. (1979). Aspects of vascular anatomy and differenciation of vascular tissues and transfer cells in vegetative nodes of wheat. *Austr. J. Bot.* 27: 703–711.

Butterfield B.G., Meylan B. (1980). *Three dimensional structure of wood. An ultrastructural approach.* (second ed.) Chapman and Hall. London.

Camefort H. (1977). *Morphologie des végétaux vasculaires.* Doin Paris.

Carlquist S. (1988). *Comparative wood anatomy.* Springer Verlag. Heidelberg.

Catesson A.M. (1964). Origine, fonctionnement et variations cytologiques saisonnières du cambium de *l'Acer pseudoplatanus* L (Acéracées). *Ann. Sci. Nat. Bot.* 1re série, V, 229–498.

Cell and Molecular Biology of wood formation (2000). Savidge R.A., Barnett J.R., Napier R. (ed.), Bios Scientific Publ. Ltd. Oxford.

Chaffey N.J. (2000). Cytosqueleton, cell walls and cambium: new insights into secondary xylem differentiation. In: *Cell and Molecular Biology of wood formation*. Bios Scient. Publ. Ltd. Oxford chap. 2: 31–42.

Cheadle V.I. (1943). The occurrence and types of vessels in the various organs of the plants in the monocotyledons. *Am. J. Bot.* 29: 441–450.

Cruiziat P. (1984). Some aspects of the water relation between different organs of the same plant. Wiss. *Z. Humbolt Univ. Berlin, math, nuturwiss. Reihe* 33 (4): 356–359.

Emberger L. (1960). *Traité de Botanique Systématique*. Tome II. Masson. Paris.

Esau K. (1965). *Plant anatomy* (2e ed.) John Wiley & Sons Inc., London.

Fahn A. (1990). *Plant anatomy* (4e ed.) Pergamon Press, Oxford.

Finck S. (1992). Transparent wood. A new approach in the functional study of wood structure. *Holzforschung*, 46: 403–408.

Fujii T. (1993a). Application of a resin casting method to wood anatomy of some japanese Fagaceae species. *IAWA J.* 14: 273–288.

Fujii T. (1993b). Application of resin casting method of wood anatomy. *Plant Morphology* 5: 3–17.

Fujii T., Hatano Y. (1996), A modified resin casting method with the application of thermoplastic resin (poster). Wood anatomy symposium. London. *IAWA J.* 17: 245.

Fujii T., Hatano Y. (2000). The LDPE resin-casting method applied to vessel characterisation. *IAWA J.* 21: 347–359.

Fujii T., Lee S.J., Kuroda N., Suzuki Y. (2001). Conductive function of intervessel pits through a growth ring boundary of *Machilus Thunbergii*. *IAWA J.* 22: 1–14.

Funada R., Furusawa O., Shibagaki M., Miura H., Miura T., Abe H., Ohtani J. (2000). The role of cytoskeleton in secondary xylem differentiation in conifers. Chap. 19: 255–262. In: *Cell and Molecular Biology of wood formation*. Bios Scient. Publ. Ltd. Oxford.

Gardner D.J., Taylor F.W. (1983). A technique for observing the exterior morphology of intact vessel conduits. *IAWA Bull.* (4) 113–117.

Gray J.D., Kolesik P., Høj P.B., Coombe B.G. (1999). Confocal measurements of the three dimensional size and shape of plant parenchyma cells in a developing fruit tissue. *Plant J.* 19: 229–236.

Grew N. (1682). *The anatomy of plant* 2e ed. W. Rawlins London.

Hartig T. (1837). Vergleichende Untersuchungen über die Organisation des Sammes der einheimischen Waldbäume. *Jahrb. Fortschr. Forstwiss. forst. Naturkunde* 1: 125–168.

Hartig T. (1854). Über die Querscheidewände zwischen den einzelnen Gliedern der Siebröhren in *Cucurbita pepo*. *Bot. Zeit.* 12: 51–54.

Hejnowicz Z., Kurczynska E.V. (1987). Occurrence of circular vessels in isolated segments of *Fraxinus excelsior*. *Physiol. Plant.* 83: 275–280.

Hossler F.E. (1997). Unusual features of the microvasculature of the urinary bladder revealed by vascular corrosion casting. In: *Recent advances in microscopy of cells tissues and organs* P.M. Motta éd. A.D.E. Med. Sc. Publ. Roma, pp. 501–506.

Kitin P., Sano Y., Funada R. (2001). Analysis of cambium and differentiating vessel elements in *Kalopanax pictus* using resin cast replicas. *IAWA J.* **22:** 15–28.

Knebel W., Schnepf E. (1991). Confocal laser scanning microscopy of fluorescently stained wood cells: a new method for three-dimensional imaging of xylem elements. *Trees* (5): 1–4.

Koek-Noorman J., Ter Welle B.J.H. (1976). The anatomy of branch abscission layers in *Perebea mollis* and *Naucleopsis guyanensis* (Moraceae). *Leiden Bot.* Ser. 3: 196–203.

Kumazawa M. (1961). Studies on the vascular course in maize plant. *Phytomorphology* 11: 128–139.

Kurczynska E.W., Hejnowicz Z. (1991). Differenciation of circular vessels in isolated segments of *Fraxinus excelsior. Physiol. Plant.* 83: 275–280.

Larson P.R. (1982). The concept of cambium. *In* Baas P. *New perspectives in Wood Anatomy.* Martinus Nijhoff Publish. The Hague.

Larson P.R. (1994). *The vascular cambium.* Springer Verlag. Heidelberg.

Launay J., 1950. Le passage des colorants entre le gui et son hôte. *CR. Acad. Sci.* Ser. D. 230: 767–769.

Lev-Yadun S., Aloni R. (1990). Vascular differenciation in branch junctions of trees: circular patterns and functional significance. *Trees* 4: 49–54.

Lev-Yadun S. (1996). Circular vessels in the secundary xylem of *Arabidopsis thaliana IAWA J.* 17: 31–35.

Malpighi M. (1686). *Opera Omnia,* R. Scott, Londres.

Martre P. (1999). Architecture hydraulique d'une talle de Fétuque élevée (*Festuca arundinacea* Schreb). Implications pour les relations entre la transpiration et l'expansion foliaire. *Thèse.* Univ. de Poitiers.

Mauseth J.D., Fujii T. (1994). Resin casting: a method for investigating apoplastic spaces. *Am. J. Bot.* 81: 104–110.

Mori S.A. (1997). *Guide to the vascular plants of Central French Guiana.* Part 1. Pteridophyts. Gymnosperms. Monocotyledons. The N-Y. Bot. Garden. ed.

Murakami T., Unehira M., Kawakami H., Kubotsu A. (1973). Osmium impregnation of methyl methacrylate vascular casts for scanning electron microscopy. *Arch. Histol. Jap.* 36: 119–124.

Nägeli C. (1858). Das Wachstum des Stammes und der Wurzel bei den Gefässpflanzen und die Anordnung der Gefässtränge in Stengel. *Beitr. Wissensch.* Bot. 1: 1–56.

Neef F. (1914). Über Zellumlagerung. Ein Beitrag zur experimentelle Anatomie *Z. Bot.* 6: 465–547.

Neef F. (1922). Über polares Wachstum von Pflanzenzellen. *Jahrb. Wissench.* Bot. 61: 205–283.

Okada S., Ohta Y. (1994). Microvascular pattern in the retina in the japanese monkey. *Scanning microsc.* 8(2): 415–427.

Okumura S., Harada H., Saiki H. (1976). *SEM observations of xylem cell walls using resin casting method* .Jap. Wood Research Soc. 26th annual meeting. Tokyo.

Patrick J.W. (1972). Vascular system of the stem of the wheat plant. I. Mature State, II. Development. *Austr. J. Bot.* 20: 49–63 and 65–78.

Percival J. (1921). *The wheat plant,* E.P. Dutton and Co. New York.

Prat R., André J.P., Mutafschiev S., Catesson A.M. (1997). Three dimensional study of the intercellular gas space in *Vigna radiata* hypocotyls. *Protoplasma* 196: 69–77.

Rhône Poulenc Silic. Dept. (1988). *Silicones. Production and applications.* Ed. Nathan Techno.

Robert D., Catsson A.M., 2000. *Biologie Végétale.* T. 2: Organisation végétative. (2e ed.) Doin, Paris.

Sachs T., Cohen D. (1982). Circular vessels and the control of vascular differentiation in plants. *Differenciation* 21: 22–26.

Saiki H., Goto T., Kakuno T. (1975). Scanning electron microscopy of glue lines separated from plywood. *Mokuzai Gokkaishi* 21: 283–288.

Sallé G., Frochot H., Andary C. (1993). Le Gui. *La Recherche* (24) 1334–1342.

Sanio K. (1873–74). Anatomie der gemeinen Kiefer. 2. Entwicklungsgeschichte der Holzzellen. *Jahrb. Wissench. Bot.* 9: 50–126.

Savidge R.A. (1996). Xylogenesis, genetic and environmental regulation. A review. *IAWA J.* 17: 269–310.

Sharman B.C. (1942). Developmental anatomy of the shoot of *Zea maïs* L. *Ann. Bot.* 6: 245–282.

Smith P.L., Gledhill D. (1983). Anatomy of the endophyt of *Viscum album* L. (Loranthaceae). *Bot. J. Linn. Soc.* 98: 29–53.

Smulski S.J. (1980). Woody cell wall penetration by water borne alkyd resin. *M.S. thesis.* Suny College of Environmental Science and Forestry. Syracuse. New York.

Soltis E.D., Soltis P.S. (2000). Contribution of plant molecular systematics to studies of molecular evolution. *Plant Mol. Biol.* 42: 45–75.

Stebbins G.L. (1974). *Flowering plants.* Harvard Univ. Press. Cambridge, Ma.

Stieber J. (1981). A new method of examining vessels, *Ann. Bot.* (48) 411–414.

Taneda K., Kawakami H., Ishida S., Ohtani J. (1979). Observations of the polymer in wood-polymer composites. I Dissolution of wood substances in WPC and shape of the polymer cast. *Mokuzai Gakkaishi* (25) 209–215.

Thoday D. (1956). Mode of union and interaction between parasite and host in the Loranthaceae I. Viscoideae, not including Phoradendreae. *Proc. roy. Soc. London. B.* 145: 531–548.

Tyree M.T. (1993). Theory of vessel-length determination: the problem of non random vessel ends. *Can. J. Bot.* (71) 297–302.

Wolfe K.B. (1962). A method of preparing mammalian lungs for anatomical study. *The Lab. Digest 25* (cité par Gardner et Taylor).

Yulong D., Liese W. (1997). Anatomical investigations on the nodes of bamboos. *In:* G.P. Chapman (ed.) *The bamboos,* Acad. Press, London, 269–283.

Zee S.Y. (1974). Distribution of vascular transfer cells in the culm nodes of bamboo. *Can. J. Bot.* 52 (2) 345–347.

Zimmermann M.H. (1983). *Xylem structure and the ascent of sap.* Springer Verlag. Heidelberg.

Index